Understanding the Environment Bridging the Disciplinary Divides

Environmental analysis and policy call on the joint efforts of many scholarly disciplines: biology, engineering, the basic physics of climate and weather, the chemistry of wastes and diffusion, and economics for measures of the impact on society and the implications of alternative policies. This book is a pioneering effort in introducing the fields one to another and presenting the elements of their aims and accomplishments so that others may appreciate and use them.

Kenneth Arrow, Professor of Economics (Emeritus), Stanford University and 1972 Nobel Laureate in Economics.

We understand the complexities of climate change because atmospheric physicists, evolutionary ecologists, and environmental economists have learned enough about each other's specialties to untangle the problem together. This text provides an introduction to how to understand other disciplines that will make this crucial step easier for the next generation of scientists.

Richard B. Norgaard, Professor of Energy and Resources Group and of Agricultural and Resource Economics, University of California at Berkeley.

Written by an impressive group of Australian National University experts arrayed across many disciplines, this book provides the reader with a coordinated introduction into each of the major elements needed to develop a sophisticated and integrated understanding of environmental issues, past, present, and future.

Kirk R. Smith, Brian and Jennifer Maxwell, Endowed Chair in Public Health School of Public Health, Environmental Health Sciences Division University of California Berkeley.

The last thirty years have seen an escalating interest in the environment, as global population burgeons and the products of human behaviour increasingly affect the biosphere. The problems are so complex that they call for a 'whole of knowledge' approach. This book of essays by experts from a wide range of disciplines makes a groundbreaking attempt to bridge the disciplinary divides so that future investigators can better achieve this holistic approach. It can be warmly recommended to university students embarking upon the very important field of environmental studies.

Frank Fenner, former Director of the John Curtin School of Medical Research and Foundation Director of the Centre for Resource and Environmental Studies at the Australian National University.

Research that seeks to address the complex interlinkages that underpin environmental sustainability requires multidisciplinary perspectives. This book exemplifies how the understanding resulting from interactions between different disciplines is greater than the sum of its parts. It will be essential reading for those who seek to improve their knowledge of contemporary environmental issues.

Andrew Haines, Professor of Public Health & Primary Care and Dean of the London School of Hygiene & Tropical Medicine.

UNDERSTANDING THE ENVIRONMENT

BRIDGING THE DISCIPLINARY DIVIDES

R. Quentin Grafton
Libby Robin
Robert J. Wasson
editors

UNSW PRESS

CAER

A UNSW Press book

Published by
University of New South Wales Press Ltd
in association with
Centre for Applied Economic Research
University of New South Wales
Sydney NSW 2052
AUSTRALIA
www.unswpress.com.au

First published in 2005

The National Library of Australia Cataloguing-in-Publication entry
Grafton, R. Quentin, 1962- .
Understanding the environment: bridging the disciplinary divides.

Bibliography.
Includes index.
ISBN 0 86840 912 X.

1. Communication in the environmental sciences.
2. Intellectual cooperation. 3. Environmental sciences – Philosophy.
4. Environmental economics – Philosophy. I. Robin, Libby, 1956- .
II. Wasson, R. (Robert), 1948-. III. Title.

333.7014

Cover design Di Quick
Cover photo Dianne Thompson, Bajabout, Canberra
Printer BPA Digital

Dedication

The editors dedicate this book to their children:
Ariana and Brecon; Kate and Billy; and Anton.

Contents

List of Figures

Foreword

This book sets out a series of essays presenting various perspectives on interdisciplinary environmental research. Such a volume, written by scholars who have practised what they preach, is particularly timely. For the last two centuries, and especially since 1950, burgeoning human enterprise has triggered environmental problems at all scales from local to global. What we are experiencing is a fundamental shift in the human-environment relationship. An equally profound shift is required in the ways that the scholarly community addresses such challenges.

The essays in this volume cut quickly to the heart of the issue. It has rapidly become a cliché to say that environmental problems require multi- or interdisciplinary solutions. *How* to achieve such interdisciplinarity in an effective way is a much more difficult issue. Critical ideas on the 'how question' are highlighted at many points throughout the volume. For example, carefully defining the questions around which the research approaches will be built is an essential pre-requisite, and the questions must be devised together by all of the scholarly communities involved. Understanding each other's concepts and languages is equally essential. However, the most profound challenge to bridging across disciplines, and the most common stumbling block that prevents the building of such bridges, is the often sharp contrast in world views. We must challenge and explore our own fundamental perceptions of our environment, our place in it, and the ways in which we generate knowledge about them. This book, more than any other I've come across, dives deeply into these profound issues.

Those who still feel comfortable in their disciplines will also find a place in the writings here. Strong disciplinary underpinnings are required to tackle complex environmental problems. However, the relationship between the more traditional disciplines and the newer integration of disciplines is often incorrectly described elsewhere as the latter being a subdiscipline of the former. This book offers a more mature perspective — that the integrative approaches now being deployed to tackle environmental problems are becoming 'grandparent disciplines'.

Bridging the disciplinary divide is a critical step, and perhaps the hardest one, towards achieving a new level of complex understandings. Theory is pushing the frontiers of what we know about complex systems in principle. For tackling real-world problems, the power of bringing disciplines together in an integrated way is indispensible. Our environment is a complex system at any scale, and the planet as a whole is the most complex subject known to humanity. Achieving an ability to understand Earth systems is the most challenging task facing scholars this century. The stakes are high. The future of Earth's life support system might ultimately depend on the knowledge base generated by the type of scholarship described in this volume.

Will Steffen
Executive Director
International Geosphere-Biosphere Programme
Stockholm, Sweden

Preface

The process we took to write this book is, in some ways, as important as the product of our labours. Unlike many edited books, this is not simply a collection of individual contributions linked by a common theme. Many of the contributors work together at an interdisciplinary environmental research centre (the oldest and largest in the Southern Hemisphere) — the Centre for Resource and Environmental Studies* (CRES) of the Australian National University, and all contributors are members of the National Institute for Environment at the same institution. We also share a strong motivation to cross the disciplinary boundaries. All of us have encountered barriers to understanding the environment due to our ignorance about domains of knowledge beyond our specialisation. We believe that our book is an important step towards overcoming this handicap.

Our motivation to write *Understanding the environment* began with the premise that barriers to communication across disciplines are hindering the exchange of mutually beneficial ideas. To overcome what we call the 'disciplinary divide', chapter contributors participated in four one-day workshops that generated a dialogue about what separates (and connects) the different bodies of environmental knowledge. This dialogue forced us to explore fruitfully our disciplinary differences and connections. What the process of writing brought out was how underlying similarities (in questions and in aspirations, if not in methods) provide a basis for making important connections across the domains of knowledge. The extent to which the chapters differ is a reflection of our tolerance of disciplinary distance.

It was not just the authors who built up the shared connections and understanding of disciplinary difference. The chapters were generously reviewed by many of CRES's other staff, visitors and disciplinary colleagues from near and far. The people who reviewed chapters of the book at various stages, and to whom we are very grateful, include Susanne Callaghan, Richard Damania, Paul Faulstich, Waveney Ford, Hannah Hueneke, Geoff Hope, Joanne Macallister, Mark McKenna, Frank Mills, Barry Newell, Alana O'Brien, Katy Penman, Mitzy Pepper, Richard Reilly, Michael Ryan, Scott Slocombe, Lucy Sneddon, Peter Somerville and Pat Werner. We also thank Bernadette Hince for her valuable copy-editing.

We also acknowledge the assistance of all of the support staff and many students at CRES for helping us to bring the book to fruition. We especially thank our colleagues Judy Clark and Pat Troy for actively encouraging the idea of the book at its earliest stages and Margo Davies for her help in organising our four workshops.

The notion that disciplines should work together, listen to each other and ultimately consider questions that transcended their original scope is a primary concern of Robert J. Wasson, director of CRES from 1999 to 2004, during which time this book was conceived, developed and written.

R. Quentin Grafton and Libby Robin
The Australian National University
Canberra

* For further details about CRES please consult http://cres.anu.edu.au/

Biographies of Contributors

Richard Baker (Geography)

Richard Baker is Reader and Geography Convenor at the School of Resources, Environment and Society at the Australian National University and has taught Geography at ANU since 1994. He was awarded the ANU's prize for teaching excellence in 1996 and 2002. His first year course 'Resources, Environment and Society' won the 2003 Australian National Teaching Prize for the best large first year course. His research focuses on Indigenous land use issues and community participation in resource management and environmental policy.

Daniel Connell (History)

Daniel Connell is a historian who has worked in the Australian Broadcasting Corporation and the Murray-Darling Basin Commission. He has published an interview-based history of life in Australia during the Second World War, produced a biography of the Labor politician Clyde Cameron and recorded many interviews for the National Library's oral history program. He is now in the final stages of a PhD at the Centre for Resource and Environmental Studies in the Australian National University investigating the role of government in the Murray-Darling Basin since the 1850s.

Steve Dovers (Public Policy)

Stephen Dovers is Senior Fellow, Centre for Resource and Environmental Studies, Australian National University, where he undertakes research and teaching in policy and institutional dimensions of sustainability, science-policy linkages, adaptive resource management and environmental history.

R. Quentin Grafton (Economics)

R. Quentin Grafton is Professor of Economics at the Asia Pacific School of Economics and Government at the Australian National University. He is the author of numerous journal articles in environmental and resource economics and several books. His latest book, *The Economics of the Environment and Natural Resources*, is published by Basil Blackwell, 2004.

Donna Hazell (Ecology)

Donna Hazell is a landscape ecologist with an interest in understanding the ecological consequences of landscape change in agricultural areas. Her main area of expertise is frog ecology. As an Australian National University postdoctoral fellow her work examined the impacts of salinity on frogs, social and ethical considerations of ecological research and policy analysis of incentive programs for conservation. Donna currently works for state government delivering on-ground biodiversity outcomes on private land.

Rob Heinsohn (Ecology)

Rob Heinsohn is an evolutionary/conservation biologist and Fellow at the Centre for Resource and Environment Studies at the Australian National University. He has conducted several long-term field studies of social vertebrates including cooperative passerine birds in south-eastern Australia, lions in Tanzania, and social parrots on Cape York Peninsula and in New Guinea.

Michael Hutchinson (Mathematical Sciences)

Michael Hutchinson is a mathematician and Professor at the Centre for Resource and Environmental Studies at the Australian National University. His principal interest is the development and application of spatial and temporal analysis techniques for biophysical data. His digital terrain and topographically dependent surface climate models are widely used for natural resource assessment and to assess potential impacts of climate change.

David Lindenmayer (Ecology)

David Lindenmayer is Professor of Conservation Biology and Landscape Ecology at the Centre for Resource and Environmental Studies, Australian National University. He has been working on various aspects of habitat fragmentation, landscape change and biodiversity response for the past 21 years. He has published widely in these fields including 13 books and over 350 scientific articles.

A.J. McMichael (Human Health)

Tony McMichael directs the National Centre for Epidemiology and Population Health, at the Australian National University. His research spans various environmental and social influences on health, including the health impacts of global environmental change. His most recent book is *Human Frontiers, Environments and Disease: Past Patterns, Uncertain Futures*, Cambridge University Press (2001). He edited *Climate Change and Human Health: Risks and Responses*, published by the United Nations (2003).

John C.V. Pezzey (Economics)

John C.V. ('Jack') Pezzey is an environmental and resource economist and a Fellow at the Centre for Resource and Environment Studies at the Australian National University. He has written many influential articles and reports on sustainable development economics and on economic instruments of environmental management, and has worked recently for the Productivity Commission and the Hewlett Foundation.

Libby Robin (History)

Libby Robin is an environmental historian at the Centre for Resource and Environmental Studies, Australian National University. Her previous workplaces include the National Museum of Australia and universities in London and Melbourne. She is author of *The Flight of the Emu* (2001), *Defending the Little Desert* (1998) and co-editor (with Tim Sherratt and Tom Griffiths) of *A Change in the Weather: Climate and Culture in Australia* (2004).

Deborah Rose (Anthropology)

Deborah Bird Rose is an anthropologist and author of numerous prize-winning books. She is a Senior Fellow at the Centre for Resource and Environmental Studies, Australian National University. Her most recent book is *Reports from a Wild Country: Ethics for Decolonisation* (UNSW Press, 2004). Her work in both scholarly and practical arenas is focused on social and ecological justice.

Robert Wasson (Earth Science)

Bob Wasson is the Deputy Vice-Chancellor Research of Charles Darwin University and formerly Director of the Centre for Resource and Environmental Studies at the Australian National University. He has major published contributions in river catchment, soil management, landscape history and climate change. He also has an abiding interest in the integrated analysis of human interactions with river catchments in India and Australia.

Ian White (Hydrology)

Ian White is Jack Beale Professor of Water Resources at the Centre for Resource and Environmental Studies at the Australian National University. His major research interests are in the prediction and measurement of the downstream impacts of land use, especially the modelling of the groundwater dynamics of salinity and waste disposal schemes, and the sustainability of water extraction from shallow groundwater systems in coastal areas.

1 Towards an Understanding of the Environment

R. Quentin Grafton and Libby Robin

1. What is this book about?

This book has three purposes:

- to explain in a concise format how a range of different disciplines inform environmental understanding
- to communicate between and beyond these disciplines, and
- to suggest ways in which connections can be made (and barriers overcome) between domains to enrich our understanding of the environment.

Generating connections across bodies (and styles) of knowledge and integrating them together (where appropriate) enables a better understanding of environment and suggests better ways to act effectively on that understanding. To use an analogy, think of the environment as an elephant and the disciplines as blind people touching the animal. Not surprisingly, the one who touched its tail will formulate a very different impression of the animal from the one who had hold of its trunk. The whole elephant is bigger than its trunk or its tail, but no one can reach right around the animal, so we must join our partial understandings to make sense of its whole. Interdisciplinarity demands knowledge of methods and world-views beyond one's own – and also a good sense of what other disciplines can and cannot offer in the way of insights into the concern at hand.

The essays in the next ten chapters examine some of the different perspectives or specialisations that inform ways to understand the environment. Their diversity is deliberate. Although disciplines other than those chosen here also constitute 'environmental studies', these explore a broad range of world views and ways of knowing. Together they form a basis for a wide range of interdisciplinary connections, of ways of going beyond a single discipline – whether that means combining two of these or all ten. Our final chapter identifies some the connections among disciplines, as well as some of their differences.

2. Communication and connection

In order to work properly, people in all domains of environmental knowledge need to communicate with many different audiences. Whether we are writing for fellow researchers, policy-makers, community leaders or simply the proverbial 'person on the street', there are some basic rules of good communication. What is communicated should be *relevant* (it must connect to the concerns of its audience), *effective* (the message must be understood by the audience in the way it was intended to be understood) and *informative* (it must provide insights that the audience did not have before).

In the environmental arena there are many 'messengers'. Although there are notable exceptions, many environmental researchers and policy-makers avoid communication with the wider public, preferring to keep within their personal comfort zones, defined by their own disciplines or domains of knowledge. Meanwhile there are thousands of environmental groups that seek to educate, protest, lobby and advocate, as well as individual researchers and policy-makers with points to make about the environment. Without clearly explained, well grounded research, many important environmental issues will be poorly understood by the broader audience of the general (voting) public.

Because the environment affects everyone, often in different ways, it is important to incorporate the perspectives of other people (and other species) in understanding how the environment works. Environmental connections are essential to confronting our environmental challenges. Ideas about how to act on knowledge must be delivered to audiences beyond any one discipline.

Accessible, good quality writing is generally far more influential than work published in prestigious journals, despite the fact that academic institutions tend to value academic output above all other. In terms of the environment, it is hard to imagine a more important book than Rachel Carson's *Silent spring*. First published in 1962, it brought new public awareness about biological processes that magnify the effects of pesticides and the long-term consequences of using them for all life forms, including humans. Carson's training as a wildlife biologist informed but did not constrain her clear writing style. Her story unfolded through case studies and metaphors that appealed to a wide audience and were quickly understood by them. DDT and chlorinated hydrocarbons became part of the vocabulary of the masses, through her metaphor for them, the 'elixirs of death'. The overarching image of a Spring without birdsong carried along many of the more complex technical details about the way chemicals behave in the land and in water systems. *Silent spring* did not just inform wildlife biology, it changed attitudes across the whole spectrum of public thinking. Popular writing like this also alerts scholars to the relevance and importance of knowledge generated beyond individual disciplines.

3. Bridging divides

Sometimes bridging divides between disciplines, such as between ecology and economics, can yield insights hitherto unknown to both parent disciplines. Indeed, the transdiscipline of ecological economics arose in response to the need to better understand the interactions between human and biological systems (see chapter 4). It has generated new ways to study environmental problems. Such combinations may have a scope so much beyond the 'parent' disciplines that they are regarded as more of a meta-discipline (or grandparent discipline). Ecology is sometimes seen as a subdiscipline within zoology or botany, but it is also an organising principle that can integrate the perspectives of both animals and plants. It sometimes reaches further and considers 'human ecology', or the ecology of societies (see chapter 7). In another example, environmental history is not just a sub-set of history, but also a means for looking historically at data generated by scientific methods, and may reach beyond human scales (see chapter 2).

Disciplinary boundaries are a particular problem when trying to understand the environment, because the environment requires knowledge from almost every discipline or domain. For instance, to comprehend (let alone predict!) the possible consequences of anthropogenic (human-induced) sources of emissions of greenhouse gases for climate change requires an understanding of meteorology, geomorphology, oceanography, botany, mathematics, economics, and many other disciplines. Obviously no single individual can hope to master every domain of knowledge, but some are keen to make connections, and do this through consulting and collaborating with other researchers with different disciplinary strengths. Some issues — like climate change — lend themselves to large multidisciplinary teams. Other issues demand only a small number of researchers, but such people need a consciousness of issues beyond the scope of their original disciplinary training. The choice of a large team or a smaller collaboration depends on the environmental issue at stake. How 'interdisciplinary' the method depends more on the style of the question than the number of disciplines represented by the team.

Understanding the environment provides ten chapters written by researchers who are recognised in their fields, and who also have been active in interdisciplinary environmental research. No special knowledge in any discipline is required to follow any of the chapters in the book. The only prerequisites are a willingness to learn, and an open mind to understand the different perspectives and ideas encountered in this book.

4. Knowledge explosion and disciplinary distance

All the scholars in this book are actively seeking to bridge gaps, to enable their discipline to contribute to a broader melting pot known as environmental studies. Each chapter is written by one or more disciplinary specialists, who reflect on how their discipline may work with other similar (or even very different) fields to gain broader perspectives. Sometimes disciplinary differences prevent us from gaining a full picture of the causes and remedies for environmental problems. Sometimes different disciplines cannot even agree on how to define the problems, or how to use the words that frame the problems.

Gaps (some would say gulfs!) have arisen from the increasing specialisation and professionalisation of different domains of knowledge, particularly in the last half century where the number of universities and higher education institutions (and students attending them) has risen dramatically. The sheer volume of material in each discipline has resulted in a 'knowledge explosion', which has then been compounded by associated technological changes. We now have the means to both observe and analyse the world in ways that would have been unimaginable, even in the mid-20th century. Increasing disciplinary complexity and sophistication has forced researchers to consider only an increasingly narrow area of knowledge if they are to follow the often rapid developments in their specialisation. For example, it was possible for one person to write a useful book describing all the marsupial research undertaken in Australia until 1970, but when asked to update this book just seven years later the author, David Ride, declared that it was impossible to do so alone because of the knowledge explosion in the intervening years.[1]

Although science, in its various guises, has yielded amazing insights about our world and our universe, we still remain profoundly ignorant about the connections across physical, biological and human worlds. For example, we still do not know (despite an enormous amount of effort and research) whether our planet can sustain a world population predicted to peak at between 8 and 9 billion (thousand million) people. The problem of 'disciplinary distance' is, in part, a function of the payoffs from specialisation. Indeed, in an academic environment, many of the rewards come from researching at the frontier of knowledge that, by necessity, requires specialisation. We do not suggest that specialisation is undesirable, but rather that specialisation *without* an understanding of the broader picture, might mean that our questions fail to effectively engage with environmental problems.

Some interdisciplinary and cross-disciplinary interactions will not only transcend knowledge barriers, but also develop new synergies through tackling different issues and thereby build new knowledge. As a first step towards such understanding, we must have some comprehension of what each body of knowledge has to offer so as to develop a meaningful dialogue. To help achieve this dialogue, all authors of chapters have addressed the same set of questions. The emphases of the different chapters are, in part, a function of the perspectives of the authors' different domains of expertise. The domains or disciplines chosen for the book — history, anthropology, economics, human health, public policy, ecology, hydrology and engineering, environmental earth sciences, mathematics and statistics, and geography — all contribute in different ways to a broader understanding of the environment. We make no pretense that these domains are comprehensive of the disciplines that contribute to environmental knowledge, but a basic understanding of these ten disciplines (and the connections between them) offers an important step forward to understanding the environment.

In considering the integration of knowledge across disciplines, all contributors have considered the following questions, at least briefly, in their chapters:

Definitions and origins of the domain of knowledge

- How do you define the discipline?
- What are its origins and its history of thought?
- What are its boundaries and limits in terms of what its practitioners consider appropriate research problems?

Concepts and philosophy of the domain of knowledge

- What is the 'view of the world' of the discipline?
- Which of its subdisciplines are relevant to environmental studies?
- What are its 'key concepts' and postulates?
- What are its controversies?
- What does the discipline *not* do in understanding the environment?

- What are its principal contributions that further our understanding of the environment?

Methods and examples of use of the domain of knowledge

- What principal methods or tools does the discipline employ to approach research problems in the environment?
- Using examples, how are these methods or tools applied?

These questions give a rough guide to reading the different chapters of the book. But the world view of the discipline or domain shapes the emphasis of authors in any given chapter.

5. A final word

Obviously this book can be read 'à la carte' by picking particular disciplines or domains that you wish to learn about. However, we encourage our readers to investigate all the domains of knowledge described in the book, especially those that you have *not* considered in relation to your personal interests in understanding the environment.

Because we aim to be concise, we have tried to keep referencing minimal, and to point to some further readings that will provide a more extended coverage of the discipline or domain and will also have fuller reading lists. As a result, most chapters reference a few classic works or recent textbooks rather than original (and sometimes obscure) published sources of the ideas or observations.

Whether or not you take our invitation to partake of the whole 'banquet' of disciplines, don't leave the table before the dessert. Don't miss chapter 12 that considers how to bridge the disciplinary divides of environmental knowledge.

2 HISTORY AND THE ENVIRONMENT

Libby Robin and Daniel Connell

1. What is history?

History is a means for dialogue between the present and the past. A good historical work embraces the past on its own terms, and communicates to a reader whose present context may be very different from that of the historical 'actors', be they people or non-human agents. History seeks ways of building real knowledge, perhaps even wisdom, not simply through information about the past, but by understanding connections between past and present.

Decisions about significance confront all of us, deluged as we are with information. Historians have access to the largest possible data sets. But more data does not always help people make better sense of the world. Sometimes the more we learn, the more we forget. History is generally a humanist craft, centrally concerned with cultural, moral, economic and political questions, but some histories allow nature (or the environment) a role as an actor in the human drama, and some even adopt timescales beyond people. History does not focus on the accrual of information but on meaningful ways of understanding it.

2. Origins

The relationship between the past and the present is something that every human culture has wrestled with. The question of how we got to be where we are now is naturally fascinating. Most cultures have 'creation' myths — significant stories about origins and the evolution of the society they now belong to. On an individual level, who we are is also very much about where we have come from.

The first historians built on these traditions, looking at the fortunes of the powerful, particularly the political leaders of the society — kings, dictators, emperors and dynasties. Patronage followed and by the seventeenth century, many European universities supported a chair of history. The eighteenth century writer Edward Gibbon's famous six volume history *The decline and fall of the Roman Empire,* marked a change of mood. No longer was history just about acquiring power, but also about how great empires can lose it. Gibbon provides a very early example of reading *against the grain,* of putting context around narrative. Among many other things, he argued against an established view that the Roman Empire was the great precursor to the even more progressive Christian Empire that followed. Gibbon figured Christianity itself as a major cause of the 'downfall' of the Roman Empire. His scepticism, intelligence and mordant sense of humour created a work that remains in print and stimulating some two centuries later.

Documentary history and professionalisation

By the 19th century, the written word was regarded as essential empirical evidence for history-making especially in the Western world. The *document* became a highly valued tool of history. It provided the discipline with a means to professionalise, by basing narrative on agreed empirical grounds that could be replicated and checked by others. One of the historians of this period was Thomas Babington Macauley, whose major work *The history of England from the accession of James III* was published between 1848 and 1855. This work was part of a shift in the emphasis of the discipline from ancient civilisations (Greeks and Romans and others) to histories of the 'modern' world.

By the early years of the 20th century professional historians found themselves specialising in different sorts of documents, and varying emphases began to put spaces in the discipline across which there was little communication. The division between ancient and modern history was one line of fracture, but it was also increasingly difficult to keep up with divergent developments in social, economic and political history, let alone the bigger canvases of 'world' (or universal) history, and the fine detail of local. Both art and science spawned their own 'histories of', sometimes also known as the history of ideas. History of space and place was sometimes local, sometimes archaeological and sometimes undertaken in a geography department under the guise of 'historical geography' (see chapter 11).

History, politics and ethics

One of the great concerns of the 1930s was 'Whig history', a prime example of which was Macauley's *History of England*. Whig history celebrated victors and ignored losers, describing progress to present day success. No-one sought the label 'Whig historian'. It was an insulting epithet for someone else's work, suggesting that they were beholden to present power structures and had failed to read the past on its own terms. In the tense interwar years of the 1930s, historians could no longer categorise those who favoured 'progress' as the winners. They were forced to deal with the ethical complexity of modernity.

Following debates leading to the abolition of slavery in the mid-19th century, human rights were articulated as a philosophically 'progressive' notion. 'Progress' and power were often at odds, and history-making developed increasing sensitivity to those oppressed by economics and power structures. German political philosopher and economist Karl Marx described history as 'economics in action' and focused on the relations between capital and labour. While his laws explaining historical change are no longer accepted, his focus on exploitation, the sources of political power and the causes of change fundamentally influenced the craft of history.

By the 1960s, political and social history was increasingly exploring the stories of poor and oppressed groups, including women and people of colour who had previously been excluded from the historical record. The ethics of relationships with non-human sentient beings (including animal rights) and the rights of nature were the next logical step, driving some of the earliest examples of the most recent wave of environmental history to ethical as well as economic considerations (Nash 1989).

The changing objects of history demanded changes in sources. The written word (or document) was often the tool of the oppressor or powerful leader. Finding the stories of slaves, other poor working people, women and non-human agents demanded reading against the grain in the document to consider its unintended effects, not just the objects of the original writer. Increasingly, official documentary sources were supplemented by oral history and memory, unofficial writings, newspapers and other media, objects and ephemera — and in the environmental area, by the landscape itself. Imagination and experience are essential tools for a historian. The British economic historian R.H. Tawney (cited by Hancock 1976: 21) declared that historians need not only more documents but also *strong boots*.

3. Methods

The central concept in history is *time*. The arrow of time moves forward, but not always in equal leaps. Perceptual time is lumpy, so scale varies between and within stories, and the art of history demands constant shifts of *scale*. Remembering is a complex process, and there is no 'best way' to do this. Each approach has its limitations, which is why a combination of approaches can be more powerful than any one alone.

Narrative

One of history's most important methods is *narrative* or story-telling. The aim of historical narrative is to build a new connection between the present and the past. It need not be an exhaustive story, but it takes evidence from multiple sources, in order to find meaning. This can be done through juxtaposition and synthesis, as well as through overt analysis. Some historians are particularly interested in theory, whilst others are drawn to imaginative understanding of events. Economic and environmental historian W.K. Hancock valued both styles. He argued that historical narrative can be based on a 'succession of propositions' or 'a succession of images' — depending on whether your preference is to know with your head (*savoir*) or with your heart (*connaître*) (Hancock 1976: 20). Historical narrative communicates with many different readers, integrates material from diverse sources and promotes discussion of the meaning of that material for human society.

Sources

Evidence for a historical narrative may come from eclectic sources — from documents, people's memories, objects, institutional structures and landscapes. The best history cross-matches different types of source, to gain a range of perspectives on the original context of action. Some historians, especially those in the area of economic history, garner statistical and other quantitative data and look at changing patterns over time (cliometrics). Others will place more emphasis on diaries, correspondence and other 'material culture' (found objects) to consider non-quantifiable, qualitative change. Among the humanities, environmental history has shown an unusual capacity to draw on material produced by the physical, earth, biological and ecological sciences and this has been an important source of its effectiveness and recent increase in popularity.

Oral history

The use of oral testimony is particularly important when dealing with Indigenous and other people who do not (necessarily) have documents. It is also valuable in shedding light on the past 'in living memory' for many other histories, but must be used with caution. An actor in a historical event may identify significant factors in a decision-making process or unexpected documentary sources, but memory is often overlaid with a present context. There is a limit to what interviews can do. An interview is often more about present views of the past than the past itself. Memory and nostalgia are not history.

Material culture (objects)

The history of ancient civilisations pieced together by archaeologists may have to be almost entirely based on found objects, because the players are no longer alive today. But historical archaeologists welcome ideas from other disciplines (for example, linguistics and geology) to enrich the context of the very imperfect material record. It is worth noting that some of the newest and most exciting archaeological digs are in the central business districts of major cities, so archaeologists are combining material culture with photographs and memories to create new urban environmental histories of the recent past.

4. What are the boundaries and limits of history?

Scientists sometimes regard history's inability to offer predictions about the future as its most serious limitation. The future is, by definition, unknown. History is an empirical craft. Whilst some disciplines can use a computer to sort and pattern large data sets, and use the languages of mathematics and statistics to smooth the bumps and extrapolate future trends from these patterns, prediction and modelling are not traditionally part of history writing. In some cases, historical sources can usefully contribute data for models crafted in other disciplines, but future-gazing is not a central aim of history. Its focus is often on the particular and the local, rather than the universal, and it defies mathematical patterning. In one sense, it is the ultimate craft of chaos. In another, it is an evidence-based craft, set in the real world.

History can ask how an outcome might have been different if a different decision had been taken. Thus, it can change variables or run scenarios, but only in relation to past, real events. Futures are in this way embedded in possible pasts, and can be handled by an imaginative extension to an empirical craft. This imaginative dimension might be particularly important in relation to certain truths, especially those that cannot be expressed mathematically, with linear (or other mathematically-definable) trajectories into the future. History embraces the non-mathematically definable, and thereby ensures that we can consider futures that cannot be projected from known present scenarios, but which might still eventuate.

5. History's 'view of the world'

History's narrative or conversation between the past and the present liberates us from assuming that the way things are now is the way they have always been (and always will be). Privileging present questions without an understanding of past context is bad history.

Equally bad (either politically naive or simply inaccessible) is a description of the past without a consciousness of present concerns and sensitivities. The dynamic between the two is the craft of history. History is always contextual. Just as the ecologist studies the 'organism and its environment', the historian studies the 'action in its context'. Without the context, the evidence is incomplete. The skill is in deciding what contextual references are essential to the story and to the past–present dialogue.

6. Environmental history: its controversies and its contributions

British historian John McKenzie has commented that environmental history has 'too many points of contact with too many other disciplines to be viewed as [a]... sub-discipline'.[2] It is not simply history nor is it environmental science in the past. One shortcoming of history for those concerned with the environment is that, traditionally, the discipline has focused its major efforts on people and nations, not nature or environmental forces. The historians who have most enthusiastically encompassed the wider forces of the non-human world have often been those who have wanted to write on timescales longer than, say, a single life or the rule of a parliament or the reign of a king. Historical scientists have long been aware of the crucial significance of temporal order in nature at many scales, but historians have only relatively lately engaged with earth scientists, palaeontologists, ecologists and others to look at 'nature in the past' alongside people and dynasties. Historical geographers have also been active in this field, and the boundaries between environmental history, geography and historical sciences are often difficult to maintain. Works like George Perkins Marsh's 1864 *Man and Nature* is often claimed as foundational 'geography' — but it is equally about history and environmental management. Marsh's idea that human forces were at work, shaping the natural world over time, for better or for worse, marked a new 'environmental consciousness' that has strong echoes for the 21st century, where change is happening even faster (Lowenthal 2003).

Most historians with an interest in the environment focus on its effect on people, rather than the other way round. Frederick Jackson

Turner, the author of a famous frontier thesis, argued that conditions on the moving western frontier in the United States were a primary force shaping American institutions and the character of its people. Karl Wittfogel speculated that the large-scale organisation needed to harness rivers and make irrigation possible was the force that created the world's first large-scale states in ancient Egypt and ancient China (Worster 1985). Similarly for Fernand Braudel (1972) and other regional historians of the French Annales School, it is the subtle but profound influence on human society of climate and geography over long timespans (*la longue durée*) that has made their histories distinctive.

More recently that emphasis has changed and there is growing interest in studying the impact of people on the environment instead of the other way round. Why should people who want to understand the environment today and in the future pay attention to the historical role of people? Because the world is a product of cultural as well as natural forces. Fire, for example, is highly cultural as Stephen Pyne (1995) demonstrates. It has had profound effects on the non-human environment. Traditional history has long had a strong focus on human activity clearly distinguished from 'nature', but environmental historians assert the interconnections between the two, locating ecological problems in the behaviour of human beings.

Many of the major controversies within the field centre on variants of the idea of 'environmental determinism'. Some biologically-trained historical writers are happy to chance their arm at describing 'big environmental forces' in human history.[3] Humanities scholars, however, are often more cautious — perhaps because of the historical links between theories of environmental determinism and Nazi policies of genocide in the 1930s and 1940s. Whilst most historians are reluctant to express simple cause-and-effect relationships between natural events, an increasing number are taking an interest in the dance between the biophysical forces of the world and the sociopolitical arenas of human life.

The dualism implicit in traditional history-making that separates people from other parts of nature, has increasingly been challenged by the greater awareness of natural forces in human life. Conversations with scientists and other environmental managers have forced historians to engage with the world in new ways that enrich the history-making craft. Conversely, natural scientists, through these conversations, are also coming to terms with human context and complexity in understanding the social embeddedness of their work, and its very specific place in time.

Environmental history embraces many styles and subjects. Urban and rural histories are treated differently, and histories of 'wild places' are different again. Histories of agricultural and other resource landscapes are dominated by economics, whilst ecological ideas (scientific, philosophical and political) permeate histories of wild places. Political and administrative institutions (such as the railway or the sewerage networks) gain prominence in urban history, but may be overlooked in other branches.

7. History in action: some case studies

Case Study 1

Chicago: a city and its ecological footprint

Environmental history has a well developed capacity to bring together genres that have traditionally been kept apart. Historians of the western United States of America, an area that has provided fertile ground for environmental history, have typically focused on the continuing westward expansion of the rural enterprise and frontier conflicts with Indigenous peoples. Urban historians, meanwhile, generally remain firmly confined within city boundaries, concerned with architecture, planning and society.

William Cronon's history of Chicago, *Nature's metropolis* goes beyond the traditional city/rural divide in a new synthesis exploring the intersections between economy and ecology. Chicago was not just a city but, in the second half of the 19th century, a force that shaped 'the landscape and the economy of the midcontinent', a metropolitan driver of the frontier itself (Cronon 1991: xv). Cronon uses commodity flows — grain, lumber and meat — and the transport links — rail and water — to trace the paths between natural systems and urban markets, to define an ecological 'footprint' of the impact of the city beyond its bounds.

Case study 2
History, environmental ideas and equity in India

Ranajit Guha (1982) critiqued the history writing of India, declaring it to be dominated by colonialist and nationalist elitism. He proposed 'subaltern studies', a pro-active project that writes histories of the missing peoples. 'Subaltern' means 'of inferior rank' and 'embraces the subordination of South Asian society whether it is expressed in terms of class, caste, age, gender or office' (Guha 1982: vii, 1). Subaltern historians typically write a great deal about agriculture and the peasant farmer, about the hill country beyond the fertile agricultural zone, and about the political organisation of people concerned about their local environments, but these studies are not predominantly 'environmental history'. They are concerned with power relations between elites and the subaltern groups. The environment is usually merely 'backdrop', not an actor in the drama.

South Asian history does not lend itself to foregrounding nature. Unlike North Americans or Australasians, some of whom actively distinguish between the natural and cultural in a landscape, South Asian landscapes are inescapably about people. This has led to a very different environmentalism, an 'environmentalism of the poor' – quite unlike the 'full-stomach' environmentalism of the wealthy countries, 'empty-belly' environmentalism confronts issues of equity and justice, in its efforts to find practices to assist a deeply damaged environment without further disadvantaging the very poor (Guha and Martinez-Alier 1997: xxi).[4] A history of nature in south Asia demands engagement with the political status of its peoples. Madhav Gadgil and Ramachandra Guha (1992) achieved this through studying political intersections between peoples and ecology across time. They dealt with the conservation measures introduced by elite groups (e.g. British Imperial foresters), but also considered 'conservation from below', efforts by people of low status to defend their own natural environments.

Although Gadgil and Guha (1992) is a short book about an ancient land with severe resource shortages and 'the most heterogeneous cultures on earth', its broad brushstroke approach suggests key themes for more detailed regional histories of India (Gadgil and Guha 1992: 2–3). For example, the controversial question of shifting cultivation (*jhum*) baffled British foresters in the 19th century, and has dogged 'modernising' Indian governments since. How do governments locate (and tax) people who depend on moving around when they own no arable land? Some scientific elites saw *jhum* as destroying ecological balance, but recent anthropological work suggests that it may be a highly efficient and ecologically sustainable use of resources, because it allows for long fallow periods. Gadgil and Guha (1992: 219) do not 'adjudicate' between Western sciences, but they summarise a variety of sources to show that *jhum* is not just an economic system with ecological effects, it is a cultural way of life. Environment and people have to be considered together, and environmental justice demands ethical as well as scientific debate.

Case study 3
History, institutions and environmental outcomes in Australia

Australia's Murray–Darling river basin provides an interesting case in the intersections between institutions and environmental outcomes. The basin extends over approximately one million square kilometres, spanning four different states and a territory, in a country where environmental responsibility is constitutionally a state responsibility. The basin's history has been shaped by the interaction between three powerful forces: landscape, climate and society (Connell 2002).

European Australians long regarded themselves as passive victims of natural forces, but in recent decades there has been a growing consciousness of the significance of human-induced change. Unintended consequences have become increasingly difficult to manage. Despite a relatively strong inter-jurisdictional river management framework, the Murray–Darling basin's natural resources are being mined, not conserved, and environmental degradation is accelerating. Future management is the subject of intense debate. The incompatibilities between culture and environment are becoming increasingly stark.

History suggests an urgent need for different scales and time frames in considering change in the basin. Within short time frames, people seem to respond to an unchanging landscape and an intractable climate, but over extended time frames it emerges that society is causing massive changes to both climate and landscape. 'Society' itself is a changing force with variable environmental impact.

Institutions have a powerful influence over environmental outcomes. When the Australian colonies federated in 1901, the political dynamic of the basin altered irrevocably. Now no one state can act alone, as other states can block objectionable initiatives. The federal system has checked some dire development options, but the collaborative arrangements in place are still failing to find effective responses to large-scale environmental concerns.

Landscape change, especially the clearing of native vegetation, has destabilised the hydrological cycle and caused the groundwater table to rise, inducing the salinisation of rivers and landscapes throughout the basin. The histories of land settlement and water management are implicated in these physical effects. Farming practices and agricultural institutions established in earlier eras, which often still shape current community expectations, need to be reconsidered. History can assist in understanding where recovery programs are beginning from, and what compensations or trade-offs might be possible. Such programs have social and ethical as well as hydrological implications.

Another major issue in the basin is the environmental flow of rivers (or water running through the system for natural, rather than for human interests.) Unlike other politically complex river systems, the original designers of Murray–Darling river management based water-sharing processes on *proportions* of available flow, rather than rights to absolute volumes. This useful legacy compensates for the extraordinary *variability* of the basin's river flows from year to year, something that is a distinctively Australian problem.[5] A good understanding of the original reasons for water-sharing arrangements, hammered out in the drought years around Federation, is essential to coping with Australia's climate variability, both now and in the future.

8. What use is environmental history?

Themes of contemporary relevance dominate environmental history, but its search for moral complexity may complicate rather than simplify the lives of policy makers and managers looking for neatly packaged 'answers'. Donald Worster (1992: 16) places environmental history squarely within the sceptical tradition of the European Enlightenment, suggesting that an environmental historian should 'question all received opinions, … take alternative views seriously, … and constantly work to demythologise the past'.

Patricia Limerick (1987), like Worster, has written extensively about the relationship between culture and environment in the American West. She creates a compelling picture of a society that failed to come to terms with multi-ethnicity, class or environment. The mining, rather than conservation of its natural resources was, she argues, a factor in its (torn) social fabric. Lacking in self-knowledge, in her view possessed by a gambling mentality and always on the lookout for the next bonanza, the peoples of the American West developed little sense of mutual obligation, espoused a narcissistic freedom and became highly litigious. Resentfully dependent on the federal government, they pictured themselves as the victims of outside forces: economic, political, social, racial and environmental. Limerick assigns to the historian the role of 'truth teller', with perhaps some echoes of Lear's fool telling the King 'truths' he does not want to hear. Yet her unflattering portrayal of the West sought to help its inhabitants understand themselves in order to move forward to a more humane, environmentally-sustainable future.

Environmental historians work within a literary, self-aware tradition of writing. Their narratives are presented as interpretations, challenging the reader, 'Is this not a convincing explanation of the facts?' In doing this they have embraced the potential of narrative in ways that make it more than a tool or medium. Environmental historians must 'advocate the distinctive skills of the story teller, to defend the logic of poetry and to champion narrative not just as a means, but a method, and a rigorous and demanding one' (Griffiths 2003: 75–6).

In environmental studies there are many disciplines that are reductive in their approach, generating insights by taking an increasingly narrow focus. The role of disciplines such as environmental history is different. They work to bring sources of knowledge together, make connections, draw out implications, and attempt to explain what all this information means for the world.

9. What does history offer environmental managers?

History is very important to interdisciplinary conversations because it uses ordinary English to trace a thread of significance through a morass of data. It eschews or tames technical terms, makes connections through story and logical argument and uses a style accessible to all. Environmental history tackles both the cultural and the scientific, and seeks broad and multi-faceted connections without losing insights at a local level. Local data are often of particular importance to management decisions.

Environmental managers might hope to mine the past for 'lessons' or ways to avoid repeating history, but this is not a primary focus of the discipline. History's chief lesson is a humbling one: human life is not ultimately amenable to reductive or predictive analysis. It can be creative and chaotic as well as cyclical or repetitive. Contingency and circumstance are as much at the heart of human history as they are in biological evolution.

History contributes most importantly when it provides a long-term view of moral dilemmas in our times, allowing us to escape the tyranny of the present. The present shapes information through 'frameworks' or context, which are always in constant flux and evolution, but appear from within the present itself to be non-negotiable – or given, fixed and certain. Historians must take sides on the moral issues that confront them, but their place is not on the judicial bench, but rather in the witness box (see Hancock 1976: 12). Seeing the world in other ways, at other times, is a way of creating a realistic alternative scenario, of framing information differently and thereby asking different questions. Unless we understand some history, we are often limited by what is currently fashionable, what is sometimes called *presentism*. A comparison with the past can throw up significant patterns in changing landscapes, or political or institutional structures, that are taken as 'givens', and allow fresh discussions, suggesting new ways forward.

10. Further reading

Environmental history is influenced by a broad literature in the *culture of science and technology* including Worster (1985, 1994) and Merchant (1980). Stine and Tarr in a bibliographical review of environmental and technological history[6] give excellent references for *urban environmental history*, another large subdiscipline. The relations between science and nature mapped out by historians of *environmental ideas and sciences* inform both history-making and scientific concepts. Glacken (1967) was a founding work in the history of environmental ideas genre. More recent works include Nash (1989), Shortland (1993) and Grove (1995). Historical sciences — including questions of when natural systems become historical — are sometimes explored through detailed vegetation and fire histories (Griffiths 2002; Pyne 1995).[7] Practical and policy applications are emphasised by some historians — for example, Dovers (2000) — whilst others explore the political dimensions of environmental sciences (Bocking 1997).

Many prominent writers in environmental history write about the United States, but the discipline has different emphases in other places. In New Zealand, the history of science, particularly pasture science, is important to understanding dramatic and very quick environmental change (see Pawson and Brooking 2002), whilst in Finland, another lightly populated, heavily forested land, ideas about national identity have predominated (see Myllyntaus and Saikku 2001). In heavily populated and environmentally poor South Africa, questions of environmental justice are foremost (see Dovers et al. 2002). In Britain, the tension between the city and the country is central to understanding environmental history and management (see Sheail 2002). Environmental histories of settler societies provide comparisons between politically similar societies in ecologically dissimilar places (for example, Griffiths and Robin 1997).

3 ANTHROPOLOGY: THE STUDY OF HUMANITY

Deborah Rose

Anthropology is defined as the study of humanity. In practice, it has developed as the study of human beings as they are embedded in social and cultural relationships. It thus differs from psychology and other studies which focus on the individual. Also in practice, anthropology developed as the study of small-scale societies, and while today it is not so circumscribed, it differs from sociology in that the methods and theories aim toward developing knowledge from analysis of small-scale particular instances and drawing broader theoretical and practical knowledge from local studies.

The main contributions of anthropology to the study of environments are focused on the relationships between people and their environments. One line of study concerns the impacts of the environment on humans, and the impacts of humans on their environments. Another more recent line of study is focused on the politics and other dynamics of relationships between humans and environments. A third (also more recent) line investigates cross-culturally some of the questions that have arisen in recent years as scholars become ever more aware of the environmental problems that the whole world is facing. This line of study asks questions such as: how do non-Western peoples manage their relationships with their environments? Can sustainability be defined cross-culturally? What can we learn from human history and from contemporary peoples, in light of the global urgency of the search for sustainable ecologies?

This chapter discusses all of these lines of study in broad overview, and focuses most particularly on the more recent questions. But first I provide an introduction to the basics, emphasising environmental issues.

1. Origins

Anthropology arose as a defined 'science' in the 19th century, and its focus at that time was primarily on describing and classifying non-Western societies and cultures. In this period many different types of human societies and cultures were being encountered through exploration, colonisation, and imperialism. The scientific emphasis was on developing a theory of social and cultural evolution within which all existing human societies and cultural forms could be accommodated. This effort sought to describe linear development from simple societies to complex societies, and from simple cultural forms to more complex cultural forms. The 19th century systems of classification were based in the first instance on a material view of society. Technology was the bottom line, and the working hypothesis was that forms of social organisation, kinship classification, and cultural production including art and religion would become more complex as technologies became more complex.

This classifying effort had the negative effect of defining some peoples as 'simple' and thus, in the thinking of the day, radically in need of development, whilst defining others as in some sense superior because of their more complex technology. It also had positive effects in showing some of the universals within the human species, and in raising questions for research about some of the important differences among human groups. In a scientific sense the work was important because the working hypothesis was proved to be wrong. Social organisation, language, religion, kinship and other cultural domains are not linked in a one-to-one correspondence with technological complexity. A good example is kinship structures. The theory of the evolution of cultural complexity was dealt a harsh blow with the discovery in the 1870s that Inuit (Eskimo) kinship structure is identical to that of modern American kinship structure. This meant that the apparently 'simple' Eskimos were socially equivalent to the apparently 'highly evolved' Americans. Ultimately, this and similar discoveries required radical rethinking of unilineal theories of evolution, and required anthropologists to develop dynamic models involving change, variability, and convergence.

2. Subdisciplines

Anthropology developed differently in the UK and in the USA, whilst still holding in common a number of core concepts. Anthropologists pride themselves on their *holism*. The 'four-field' identity of anthropology includes linguistics, social/cultural anthropology, archaeology, and physical or biological anthropology.

Linguistics is the study of human languages. Archaeology is the study of past human societies based primarily on their physical remains. Biological or physical anthropology is the study of past and present human biology. An important component is primatology and the development of the anatomically modern human (*Homo sapiens sapiens*), our physical history, geographical diffusion, and contemporary variability. Biological or physical anthropology also includes contemporary studies of human biology as it is influenced by or influences society and culture. Social and cultural anthropology are concerned with contemporary peoples, and focus respectively on society and culture (see section 4, below).

3. World view

The anthropological world view starts with the biological fact that humans are social animals. Humans must be socialised to participate in society, and must learn many domains of culture for both personal and group survival. Language, different types of knowledge and practice, rules and goals are among the cultural domains which universally must be learned. The second fact is social: societies socialise their members to engage in productive social roles. Societies sustain themselves across generations, so there is a process of social reproduction that must achieve some degree of adaptive fitness in order to persist.

In respect of environmental issues, there is broad consensus around the idea that the era in which our species emerged (the Pleistocene, lasting from approximately 1.7 million years ago to 10 000 years ago) was marked by great climatic instability. There were ice ages, changes in sea level, and many terrestrial changes associated with these large climatic fluctuations. Our species 'has specialized in being an ecological generalist, consuming an amazing variety of foods, inhabiting an unprecedented range of habitats and climates, and developing a wide range of very flexible and opportunistic niches' (Smith and Wishnie 2000: 497, citing Gamble 1994). As a flexible, generalised, and opportunistic species, we quickly occupied every terrestrial environment on earth that supports mammalian life.

4. Key concepts

Holism

Anthropology's commitment to *holism* in the study of human beings includes the four fields, and thus spans the natural and social sciences and the humanities. Anthropologists specialise, and although no professional is expected to be equally competent in all four fields, university courses in the four fields make up part of the standard undergraduate and graduate curricula.

Culture

Within this broad holistic frame, the key concept is *culture*. Culture is defined as a system of knowledge more or less shared by members of a society. The thinking behind the centrality of the concept of culture is that our physical evolution has been radically overtaken by our capacity for symbolic thought: for speech, memory, transmission of learned knowledge, and other forms of cultural action which enable us not only to escape the limits of physical evolution, but also significantly to shape the world in which we live through action based on knowledge and imagination. A frequently cited example is that when we moved into cold climates we did not grow fur, but rather made clothing out of other animals. Our imagination and intelligence enable us to develop knowledge that we draw on to respond rapidly and flexibly to biological and environmental imperatives. The concept of culture does not imply that humans are not physical beings, or that they are not subject to the same requirements for survival as are other living things. It does imply that humans shape their understandings of their experience in cultural forms (through language, for example), and depend on cultural knowledge to articulate and attempt to satisfy their many biological, social and cultural needs.

Society

Society is defined as a population marked by relative separation (physical or cultural) from surrounding populations, and by a distinctive culture. A society has a structure of positions and roles, and societies are systematic in maintaining structures of social relations over time.

Cultural relativism

American anthropology in particular has emphasised *cultural relativism*. Cultural relativism is an ethical and scientific position that insists that since cultures are diverse and unique, they must be understood in the first instance in terms of their own standards and values. It must be emphasised that cultural relativism does not require abandoning one's own ethical standards. Cultural relativism can be carried to extremes, and it is important to recognise that it arose as a key concept in the interests of an impartial analysis of society and culture.

The great merit of concepts of relativism has been to free anthropology from the ethnocentric assumptions of 19th century typologising. One of anthropology's achievements, for example, was to disprove the idea that there are essential links between race, language and culture. Anthropology generally has held a strong line against racism and racially-based theories, and continues to do so. The open-minded inquiry demanded by cultural relativism positions us to make comparisons based not on our own prejudices but on evidence acquired through detailed open-ended research into the varied ways of being human.

Ethnography

An *ethnography* is an account of a way of life. The rise of anthropology as a field discipline was closely linked to the rise of the participant observation method discussed below (see section 5). Ethnography is a product of anthropological research, and is fundamental to the distinctive anthropological way of learning and knowing. The best evidence for or against a given theory is generally held to be ethnographic evidence (discussed further in section 8).

5. Methods

Armchair anthropology

The early methods in anthropology were related to the classifying endeavour, and they are now termed *'armchair anthropology'*. Through the use of questionnaires and detailed correspondence, armchair anthropologists developed a data base drawing on information from colonisers, police, missionaries, travelers, settlers, and various other colonial frontier opportunists. They also trawled the myths, legends and folklore of their own societies, seeking correspondences among different European traditions and with contemporary peoples elsewhere in the world. The quality of the early questionnaire information is almost invariably suspect, but even in its inadequacies it was sufficient to disprove many 19th century cultural assumptions (as discussed above).

Participant observation

Participant observation seeks to overcome problems of quality of evidence associated with earlier armchair anthropology. Developed in the first two decades of the 20th century (in America by Boas and in the UK by Malinowski), participant observation requires the researcher to spend time with people doing what they do, learning to learn as they learn, and gaining both the experiential knowledge of being in the situation, and the opportunistic knowledge that often cannot be formally elicited but that arises in the course of daily life. The range of methods used in the field has grown, but the hallmark of anthropological research is local, face-to-face learning.

Participant observation has had the effect of dismantling the subject/object distinction, replacing it with intersubjectivity. Anthropologists adhere to standards of objectivity; these standards do not depend on treating people as objects, however, but rather involve impartiality, accumulation of multiple perspectives, development of trust, and comparison between domains of evidence (how people's accounts of themselves relate to their actual behaviour, for example).

Within the general frame of participant observation a range of qualitative and quantitative methods are used. *Formal and informal interviews* are key methods. Other methods are designed for investigation of particular questions. For example, where land is a focus, *mapping* (see chapter 11) will be a key method, and where the land is imbued with sacred significance, geographical mapping goes hand in hand with questions of religion and cosmology. Land always exists within systems of tenure and use-rights, so mapping the social relations amongst people (*genealogies* and other forms of association) will be important. *Histories* of use-rights and investigation into *decision-making processes* (see chapter 2) will also be important.

Extended case method

Given the long-term nature of most human endeavours, a particularly important research method is the *extended case method.* The study of an event that is spread out over years, decades, or even generations, enables the anthropologist to acquire understanding of long-term processes of social and cultural life.

Ethnoecology

An important research method pertinent to environmental questions is the elicitation of local taxonomies of plants and animals. Originally known as *ethnoscience*, the research methods were developed by linguistics and were designed to elicit local (often termed 'folk') classifications. The investigation of classifications has led to a number of subdisciplines prefixed with 'ethno'. Thus, for example, ethnobotany, ethnobiology, ethnopharmacology, ethnohydrology, and others are the results of methods aimed at elucidating 'folk' knowledge of the natural world.

Ethnoecology is especially pertinent today. In its narrow sense, ethnoecology seeks a general correspondence between Western ecology and local knowledge. In its broader scope it looks to relationships and thus refuses to draw Western boundaries across local systems of connection. This enlarged scope is likely to include human-animal relations, ethics, and world view (see for example Scott 1996). There is a practical aspect to this subdiscipline. Local perceptions of nature, as expressed through social and cultural processes, potentially enrich our collective environmental understanding by providing regional specificity to global issues. Ethnoecological research looks toward a range of issues, including environmental histories, the search for new medicines, the documentation of species that are rapidly becoming extinct, and collaborative work between scientists and local peoples to develop new systems of environmental management.

Quantitative methods

Quantitative methods include measuring details of local practices such as the use of fire in hunting. They may also include large-scale surveys as appropriate. Depending on the numbers of people and variables involved, statistical analysis enables a precise formulation of the relationships between the variables, and can determine levels of significance. Complex mathematical modelling has also been undertaken to test anthropological theories (Anderies 1998).

The comparative method

The comparative method is anthropology's answer to the fact that it is neither possible nor desirable to conduct experiments with human societies. The comparative method treats the world as an enormous unplanned experiment in human diversity, reflecting adaptability, contingency and necessity. Through comparisons between different peoples, places, and times, anthropologists seek to discover the nature of human constants and variables in the diverse relationships between people and environments. Without comparison, the incredibly rich variability of the human experience would be simply documented as a fascinating play of freedom and necessity across unique fields of time and space.

In our postcolonial and decolonising world, environmental issues are requiring ever more analysis and action. At this time comparison often takes the form of dialogue. For many, comparison today is not so much an objective study from the outside, but more a politics of advocacy, persuasion, and imagination, argued from within our multiple perspectives in a crumbling global order. It is a search for common ground, knowledge sharing, and moral action.

6. Environmental anthropology

The term 'environmental anthropology' covers anthropology's efforts to include environments within the study of humanity. Within the history of anthropology, it is possible to discern a major shift in thinking. Earlier theories and methods sought to differentiate and to define single causes; more recent theories and methods seek to integrate and to look for more complex non-linear relationships.

'The environment'

In the earlier literature, the question of humans and environments sought a linear causality: is human society and culture the product of the environment? Some of the most interesting and enduring efforts to analyse the relationship between humans and environments arise in the context of the study of ancient states. State formations are identified by a set of diagnostic criteria that include social hierarchy, centralised authority, centralised distribution of goods and services, division of labour by class or caste, writing, taxation, conscription, and other familiar criteria. The original states arose in conjunction with irrigation agriculture. This strand of study is associated with the influential work of Karl Wittfogel (writing in the 1930s, 1940s, 1950s) and those who elaborated his hydraulic theory of state formation. The analytic logic has been that large-scale irrigation requires centralised control, that centralised control requires centralised authority, that authority requires a centralised economy which itself requires methods of record-keeping, taxation, and conscription, and which maximise the authority's access to human labour through elaborate and specialised division of labour and forms of coercion.

Scholars of ancient civilisations have also sought to assess the causes of collapse of these states, and it is clear that many of the causes are environmental. Salinity is a good example: the collapse of subsistence systems undercut the ability of states to sustain themselves, hence their 'collapse' back into less complex forms of subsistence and social organisation. While the connections between the control of resources and the control of society has been well studied, the lessons to be learned from ancient states concerning the absolute requirement of environmental sustainability is often ignored. Theories of evolutionary 'progress' may encourage people to ignore the lessons of the past. The theory of progress rests on an often unstated principle that humans are gaining ever more knowledgeable mastery over their environments, and that our increasing knowledge will save us from the effects of our unsustainable practices. The theory of 'progress' is just a theory, and it is now being tested on a global scale with scant consideration of what the consequences will be if, as seems likely, the theory proves to be wrong.

Cultural ecology

Cultural ecology had its origins in the classification endeavour, and for decades was concerned primarily with categorising human subsistence strategies in relation to environments. Built on possibilism rather than determinism, cultural ecology sought patterns between human social organisation, technology, and environments. To a large degree, the categories that describe these patterns are now taken for granted. They include hunting/gathering, pastoral nomadism, various forms of agriculture from swidden cultivation to intensive forms of cultivation, and industrial agriculture.

Ecological anthropology

Cultural ecology has now developed into a research focus that is often labelled *ecological anthropology*. This subdiscipline shares with the other ecologies discussed below an emphasis on analysis of the dynamic relationships between people and ecosystems. Ecological anthropology is greatly influenced by the work of Gregory Bateson in cybernetics, as well as by scientific advances in the ecology of natural systems and by advances in concepts of non-linear dynamics.

The best early example is that of the great ecological anthropologist Roy Rappaport. His 1968 study of a Papua New Guinea swidden subsistence society, *Pigs for the ancestors,* inspired a generation of anthropologists and continues to generate debate and analysis. Rappaport analysed a dynamic system of environmental adaptation in which 'ritual cycles ... play an important part in regulating the relations of these groups with both the non-human components of their immediate environments and the human components of their less immediate environments' (Rappaport 1968: 182; see also Anderies 1998). Among Rappaport's many contributions to ecological anthropology, what stands out is his emphasis on dynamic systems of feedback and action between society, culture, and environments. His work does not determine in advance which aspects of society and culture are most relevant to environmental questions, and thus he is able to include religion, ritual and other matters that are generally excluded from Western environmental sciences.

An excellent recent example of ecological anthropology is Lieber's (1994) study of fishing on a Polynesian atoll. Lieber states that Kapinga fishermen were technologically capable of eradicating any number of fish species long before colonial contact. The evidence is that they did not do so, and yet since contact they have begun to fish in unsustainable ways. Like Rappaport, Lieber brings a very open ethnographic analysis to the question of why fishing has changed. He finds that pre-contact Kapinga fishing involved 'a relatively simple technology that is complexly calibrated' (Lieber 1994: 54). The results of this complex study show that the interactions of Kapinga people and outsiders — missionaries, colonial agents, and others — destabilised the social organisation that had distributed fishermen's attention across a broad range of fish, focusing it on a few key species. These were the species that were being utilised unsustainably.

This study exemplifies Bateson's assessment of the paradoxical situation in which humans, as intentional beings, are enmeshed. According to Little (1999: 259) Bateson was concerned with the dynamics of the interplay of 'conscious purpose, which tends to be linear, and the circularity of cybernetic and biophysical systems'. Little went on to state that Bateson 'found that this interaction produces neither predictability nor randomness but rather stochastic processes within which both random and selective forces are operating.'

7. Postdisciplinary ecologies

Political ecology

Political ecology is a recent development arising out of dialogue between anthropology, geography and other disciplines. It examines connections between political economy and human environmental impacts. The central insight of the older hydrological theory of civilisation — that control of resources is readily transformed into control of political and economic systems — forms an important theoretical component to the field of political ecology. The focus is on analysis of the increasing number of conflicts centred on environmental resources. Political ecology offers social and cultural analyses of resource wars, the politics and problematics of refugees from environmental disasters, and the politics of resource management.

There are many excellent contemporary analyses within the broad scope of political ecology. For example, the study of relationships between water and power has been brought back into contemporary analysis. Worster's historical study of the role of irrigation in the settlement and political economy of the American West is a key example. Other examples include the essays collected in the major work *Water, culture and power* (Donahue and Johnston 1998). In this volume, for example, Donahue analyses water wars in south Texas, describing the many antagonists in the battle for managerial control of the Edwards aquifer. He shows how different groups framed and expressed their interests, analysing some of the covert cultural frameworks within which different positions were embedded.

Human ecology

Human ecology is another recent postdisciplinary ecological mode of analysis, closely related to ecological anthropology and including disciplines such as psychology, biology and sociology. Focusing on the interactions between people and natural systems, human ecology is largely concerned with how our species adapts to the environment. Studies in human ecology range from detailed studies of local interactions between society and environment to theoretical analysis of the connections between human biology and the environmental context of human life. Human ecologists strive to formulate theories about how people perceive, organise knowledge about, and then act upon the environment. For example, Hornborg (2001) has developed an analysis connecting recent theories of modernity with aspects of physics, cybernetics and ecology to provide a rich contribution to the understanding of contemporary environmental problems. He contends that an ideal approach to human ecology would recognise the recursive relations between the ecological, sociological, and cultural dimensions of human life.

Social ecology

Social ecology tends to draw extensively on sociology, and is particularly associated with the work of Murray Bookchin. This approach analyses the technologies and political and social institutions that people use to mediate between themselves and non-human nature. It pushes social systems toward new patterns of production, and offers theories that explain the social causes of environmental problems and alternative ways to address them.

8. Sustainability

A major debate within and beyond anthropology concerns the question of indigenous knowledge and practice in relation to ecological sustainability. The debate is often organised around the concept of environmental conservation. In this final section I discuss the terms of the debate and offer some directions for future approaches to the analysis of sustainability.

The question of the cross-cultural applicability of contemporary Western concepts of conservation has an empirical basis: it is demonstrably the case that many pre-industrial societies lived in a state of sustainable balance with their ecosystems. The question thus arises as to whether they achieved this sustainability through conscious design or whether it was simply a byproduct of other factors such as low population densities and less invasive technologies.

The problem of posing the questions within the framework of Western conservation is the old problem of ethnocentrism and cultural relativism. As Smith and Wishnie (2000: 515) note, the Western approach to conservation 'is dependent on a set of political, economic, and intellectual realities with few parallels in human history'.

They then proceed to define conservation in a highly restricted mode: to qualify as conservation a practice must involve some level of restraint or sacrifice in the present in order to achieve future benefits.

There are two main issues here. First, if the contemporary conservation ethos is unique to Western post-industrial societies, then it would be perverse to seek to discover or impose it on others. The definition already requires that people think like Westerners in order to be counted among the humans who are concerned for the preservation of ecosystems. Second, if the assertion is that contemporary Western culture is the only culture in human history that seeks sustainability, then the evidence of the vast majority of humanity must be excluded. Not surprisingly, such restrictive views of conservation, and environmentalism more generally, have been the subject of broad critique from both inside and outside the West. Many Indigenous people have suffered both genocide and ecocide, to use Grinde and Johansen's (1995) confronting terminology, and they find such views reprehensible.

What about asking people who have lived sustainably what they have to say about their own practices within their living systems? This is the basis of scientific even-handedness: to examine other cultures on their own terms in the first instance. Such an approach goes beyond the ethnocentric definitions of conservation, and holds the door open for dialogue around issues of sustainability.

In an as yet unpublished paper, Langdon contends that there is a further problem here: that of imposing a simple model of restraint on complex systems. His example concerns hunting. As is well known, restraint is not always a virtue in human engagements with the population dynamics of other species. There may be times when it is best to hunt heavily and other times when it is best to hunt lightly. A single rule that calls for restraint and sacrifice is not always the best way to engender sustainable relationships between human and other populations.

In my research with Aboriginal people of Australia, I have found elements of restraint, but they are not organised around Western preoccupations with conservation. Some forms of restraint randomise hunter/prey interactions, while others are more complex. In *Coun-*

try of the heart (Rose 2002) I offer numerous examples of people's work to enhance ecosystem viability for a whole range of species including humans, and examples of their shifts from one subsistence resource to another as a response to a set of indicators which capture large sets of information about the condition of country, resources, and other species. More significantly, perhaps, I have found that the contemporary idea that humans have to make sacrifices for the long-term good of the ecosystem is a fundamental misconception of the concept of 'interest'. The Indigenous people with whom I study are enmeshed in relationships of connectivity with other species, and in systems of knowledge and practices of care that for the most part do away with the need to distinguish between self-interest and the interests of others. Many of their practices, such as burning the country, are oriented toward making ecosystems 'good for everyone'. The 'good' that they achieve contributes greatly to their own subsistence, while benefiting numerous other species as well.

Some of the passion behind these debates is political. If Indigenous people are deemed to have achieved sustainable relations within their ecosystems primarily by accident, then it follows that they have nothing in particular to offer in current debates about the future management of ecosystems, although some groups have achieved the political right to be included in environmental decision-making. In contrast, if Indigenous people are deemed to have developed a body of knowledge and a set of practices that have enhanced their ability to live sustainably, then it is of the first importance that they be included in dialogue about ethics, practices, and possible futures not only as a matter of right but also as a matter of enhancing globally the possibilities for sustainable ecosystems.

A further aspect of the passion behind these debates concerns the essential nature of humanity. If it is found to be the case that no human groups have ever lived sustainably by design, then it may be that it is built into our nature to be destructive. It follows that those who are today at the forefront of the conservation movement (Western scientists in particular) must take a leading role in shaping the parameters of future relationships between society and environment. On the other hand, if it is the case that many human groups have lived sustainably because of planned interest (an interest mutually entangled with the interests of other living things and with life support systems) or as a result of awareness of feedback loops that enhance their long-term subsistence and survival, then it follows that our human nature is flexible. It would thus be the case that many possibilities are open to us, and that a good way forward is to engage in broad dialogue among experts in a large range of knowledge systems. I hold the view that human nature is flexible, and therefore I believe that creating ways in which such dialogue could be made workable is one of the great challenges of the 21st century.

Little (1999) concludes his essay on environments and anthropological research with a call that resonates well with what many Indigenous people and many anthropologists around the world now contend: there is an urgent 'need to develop a new attitude of caring for the earth and its inhabitants, human and other'. His reference to attitudes brings us back to the paradox articulated thirty years ago by Bateson: that motivation is usually linear while the ecosystems toward which we seek to motivate ourselves are non-linear. For scientific reasons as much as for moral and survival reasons, research may be telling us to step outside our traditional categories and to think more deeply and creatively about shifting our diffuse attitudes and desires, as well as our knowledge and practice, toward long-term interests of life on earth.

Acknowledgements

Many thanks to Paul Faulstich for his assistance with this chapter.

9. Further reading

Detailed discussions of key concepts and methods can be found in any good introductory text; see for example, Keesing and Strathern (1998). Anthropology's achievement in separating race, language and culture is first argued forcibly in Boas' 1940 *Race, language and culture*. On ethnoscience, see Berlin (1992). For more information on the impacts of Roy Rappaport's life work, see Messer and Lambek (2001). Gregory Bateson's work is perhaps most accessible in his 1979 study *Mind and nature*. Recent studies in ethnoecology can be found in Nazarea (1999). A good introduction to the work of Karl Wittfogel is provided by Donald Worster. On political ecology, see the *Journal of Political Ecology*, and on human ecology see the journal *Human Ecology*. Recent perspective on Indigenous ecologies are well expressed by geographer Fikret Berkes (1999).

4 ECONOMICS OF THE ENVIRONMENT

R. Quentin Grafton and John C.V. Pezzey

1. Nature and origins of economics

Nature

Since our earliest ancestors, people have been making economic decisions. These decisions involve *choices and trade-offs* as to how to best use time and resources among competing alternatives. For a Palaeolithic hunter-gatherer the choice could have been between hunting for a deer, fishing for salmon or searching for edible roots; while for an Egyptian Pharaoh it may have involved choices over spending on the standing army, building a temple or increasing grain storage facilities.

Today our choices differ, but the need remains to consider alternatives. These include both everyday choices like buying goods (such as bread) or services (such as a haircut) from a shop, and life choices such as what job to take or where to live. Regardless of who we are, we all face *scarcity*: that is, constraints or limits on how many desirable things in life we can have, because of limits to our time, our current income, our accumulated wealth, etc. Economics, in its most basic form, is the study of how – given these constraints – people make their choices, allocate their resources, and distribute the results.

Most of what economists study is the *quantitative* trade-offs of commodities (here, this refers to goods and services, although it also includes natural resources like coal) that people *exchange* with each other, usually indirectly through markets. Economics seeks numerical measures of the goods and services which people value, and the means of producing them, using units like hours of labour or megalitres of water. It also presumes that people make their choices to maximise their wellbeing or 'utility'.

Mainstream economics is much more concerned about the goods and services that money normally measures than about the environment. A key role for the economics of the environment is to try to value environmental services often overlooked or undervalued by the market place. In so doing, environmental economics helps ensure that the part nature plays in providing us with sustenance is given proper consideration. It also takes it a step further by analysing the ways human activities and actions can help address the use (and abuse) of the environment in ways that improve environmental quality and wellbeing.

Many introductions to economics ignore time or place as factors that affect our choices. Surprisingly, given that choices frequently involve moral dilemmas and ethical questions, superficial treatments of economics also fail to consider equity issues at either an individual or societal level. Thus, students enrolled in their first course in economics may mistakenly believe that all economics is ahistorical, ageographical and amoral. The reality is that although these issues are not given the attention they deserve, as far back as Adam Smith in the 18th century, some economists have taken a broad or holistic perspective to analyse and explain human behaviour and outcomes. To help understand what is economics, and what it offers to understanding the world around us, we will briefly explore its origins.

Origins

Economics as a separate discipline usually traces its roots back to the work of Adam Smith and, in particular, his classic text *An inquiry into the nature and causes of the wealth of nations,* first published in 1776. Although this mammoth work included digressions into the value of silver over four centuries and a history of education in the Middle Ages, the basic question addressed is clear from the title: How had some nations become so much wealthier than others? Investigating this led Smith to study individual incentives, prices and wages, property rights, social institutions, trade and economic growth. Although he was by no means the first writer to address these issues, he was the first to put them all together and attempt a comprehensive prescription of how to promote wellbeing. Many of his ideas were subsequently developed and extended by others and form the basis of most of the economics used today.

One of Adam Smith's best known ideas is the *division of labour*: that by dividing a given piece work into many tasks, each done by a separate specialist labourer, the productivity of a given number of people is often increased, through skills acquired by greater repetition. Another is that markets give benevolent guidance to groups of purely self-interested individuals and may be said to be led by an *invisible hand* to promote society's interest (Smith 1776, book IV, chapter 2).

For a century or so after Smith, the main focus of 'classical economists' like Malthus, Ricardo and Marx was the overall development and growth of nations. Central to this was the question rejuvenated in the 1980s by the debate on *sustainability*: can prosperity increase forever, or are there limits? In answering this, classical economists paid much attention to natural resources, particularly agricultural land – most famously in Malthus' pessimism (thus far misplaced) that population growth would be bound to outstrip food production.

In the late 19th century, attention shifted to individual markets within the economy, known as *microeconomics*, rather than the economy as a whole, known as *macroeconomics*. An intrinsic part of this shift was the discarding of any special role for environmental resources, and the development of *marginal analysis* that focuses on how a 'marginal' unit of consumption, production or other activities – meaning the last or extra unit consumed or produced – affects individual economic welfare. Marginal analysis uses mathematics to model individuals' (agents') actions in an economy as being subject to precisely defined laws of supply and demand, much as Newton modelled a planet's motion in a solar system as subject to precisely defined laws of gravity. This crystallised in the *neoclassical* approach to economics that has become mainstream since the 1940s, and which emphasises the benefits of markets and competition to help achieve Adam Smith's 'invisible hand' result.

Marginal analysis

Marginal analysis provides a framework for explaining a great deal of economic behaviour, as people often act as if they maximise their wellbeing ('utility') by consuming or producing up until the point where the marginal benefit of the last unit equals the marginal cost of consuming or producing it. For example, a firm would not ordinarily choose to produce a good beyond the point that its marginal benefit (usually the price it receives for selling each unit of the good) exceeds the marginal cost of producing it. This can be visualised in figure 4.1 for the case of an owner-operator firm (supplier), competing with many similar suppliers to sell an identical good to a single consumer (demander).

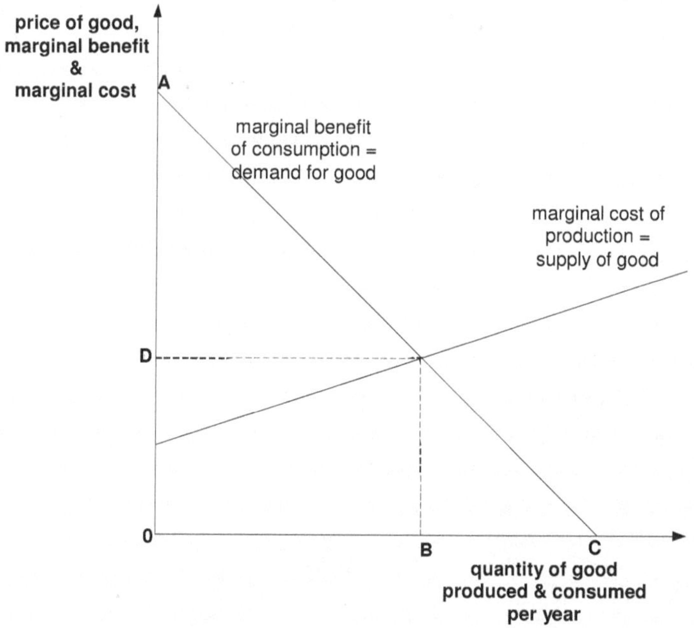

Figure 4.1. Marginal analysis and equilibrium.

The price that consumers are willing to pay for one unit of the good falls as the quantity of the good produced rises, because each additional (marginal) unit gives less benefit or value to consumers, and hence is worth a lower price. So the 'demand curve' slopes downwards, with the maximum demand price that the consumer is willing to pay being $A per unit when no units are being consumed, and with the good having no marginal benefit once C units are being consumed. By contrast, the 'supply curve' or the marginal cost of supplying (producing) the good rises as the amount produced rises. This is because the owner-operator places an increasing value on leisure time, and the time available for leisure decreases, the longer the hours that the owner works.

The preferred level of production for the firm is point *B* where the marginal benefit (i.e. price) of production exactly equals its marginal cost. Quantity *B* and its associated price *D* represent a demand-supply equilibrium for the firm because the owner has no desire to change the amount produced. Because prices are determined by demand and supply for the marginal unit, they do not reflect the total value of all consumption of the goods in question. For example, the *total* value of all freshwater resources consumed is infinitely greater than the value of all diamonds consumed; but because water is so much more plentiful than diamonds, its price (marginal value) *per kilogram* is much lower.

2. Economics of the environment

The economics of the environment as a subdiscipline has developed within only the last hundred years. A landmark work by Pigou (1920) introduced the concept of a *technological externality,* whereby a decision-maker acts in a way that either improves or harms the well-being of others, but these effects are neither considered by the decision-maker, nor 'priced' or accounted for in the market place. Thus a factory emitting polluting wastes into a river that harm the health of residents downstream is causing a negative technological externality.

Pigou's suggestion to remedy or *internalise* the externality, and reduce the excessive level of pollution it causes, was to tax or charge the factory an amount per unit of its emissions equal to the harm that an extra unit of emissions causes to downstream users. In principle, if the per unit pollution tax were set at the correct amount and equal to this marginal level of harm to downstream users, the 'efficient' level of water pollution would result. Efficient refers here to a level of pollution where the marginal willingness of users to pay for *reduced* emissions exactly equals the marginal willingness to pay of the firm to increase its emissions. But an efficient level of pollution may be very inequitable, as it fails to consider *who* bears the costs (or enjoys the benefits) of actions that affect others. Many economists, however, avoid debates about equity, and often focus solely on achieving efficiency.

Externalities

The negative externality considered by Pigou and the 'efficient' level of pollution can be visualised in figure 4.2.

The production or output of the firm (but not its emissions) in figure 4.2 generates value to others (it could, for example, be producing hospital syringes), but each unit of output also produces a fixed amount of emissions. When seeking an efficient level of pollution, we seek a balance between the value of the output to society, and its total cost to society. The value of this output to those consuming it is represented by its price, which equals the *marginal private benefit*. If the price of the product also equals its value to society, then the price will also be its *marginal social benefit*. We would expect that the more is produced and consumed, the less hospitals would be willing to pay for the product, as there are only so many syringes that even a hospital can use. So the marginal private benefit falls as the quantity of output produced and sold rises, again giving a downward-sloping demand curve.

Also as in figure 4.1, the marginal cost of production, defined as the *marginal private cost* of the output, is rising in figure 4.2; that is, the supply curve is upward-sloping. Here this is because the greater use of machinery or higher labour hours imposes higher per unit costs in the form of more regular machine maintenance or overtime rates for workers. Each unit of production, however, results in a cost for downstream users that is defined as the *marginal external cost*, so-called because it is a cost that the firm normally treats as 'external' to its financial calculations. The sum of the marginal private cost and the marginal external cost equals the *marginal social cost* per unit of output.

In the absence of a pollution tax, or any other measure to control pollution, the firm would choose to produce q' units, at the privately efficient quantity of output where its marginal private benefits equal its marginal private costs of production. From society's perspective, however, the preferred level of output is q^*, the socially efficient point where marginal social benefit exactly equals the marginal social cost of production.

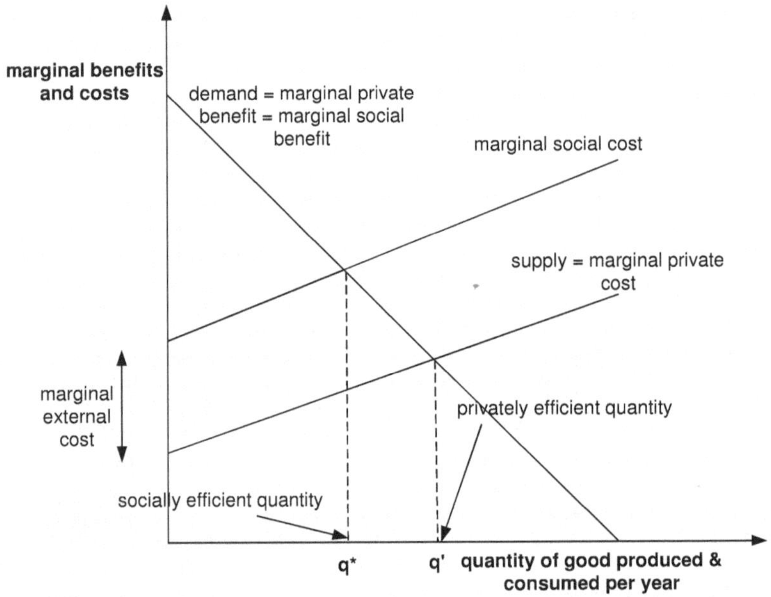

Figure 4.2. Privately and socially efficient levels of production with a negative technological externality.

Measuring marginal external costs in monetary units, so as to know where q^* might be, is an important part of the economics of the environment. Another part is designing mechanisms or institutions to move from the socially undesirable (inefficient) point q' to the socially efficient point q^*. Despite Pigou's suggestion of using emission taxes, governments have mostly opted for regulations or controls that simply prohibit firms from polluting beyond a fixed amount. More recently, the idea has become popular that tradeable property rights (say, equal to a level of emissions that corresponds to output level q^*) can be created for pollution. The price paid for such rights to pollute would be a cost to polluting firms that gives them a financial incentive to reduce emissions. Both taxes and tradeable rights amount to someone (the government, and/or private individuals) effectively 'owning' (commoditising) an environmental resource that was previously unmanaged. In principle, such *market mechanisms* or *economic instruments* of environmental management can increase *cost-effectiveness*, by reducing the overall costs of any targeted reduction in pollution, even when marginal external cost and hence the socially efficient pollution level are unknown.

Resource economics and ecological economics

The economics of the environment goes far beyond the issues of pollution control, and encompasses the challenges of natural resource management (such as the economically desirable level of mining or harvesting) commonly called *resource economics*; sustainable development (such as the choices we all need to make to ensure a desired level of wellbeing for the future); and non-market valuation (such as the value we can ascribe to a nature reserve that is provided free to the public, or the values of environmental assets to people who do not physically use them). A common characteristic of all these approaches is that most analysis focuses on the present and the future, but not the past. Understanding the history of how we arrived at the current situation or circumstances is often either ignored, or considered as very much of second-order importance by economists.

To be fully comprehensive, the economics of the environment should include some more awkward aspects, such as threshold effects (sudden, maybe, irreversible jumps) in nature, possible limits to substitutability between manufactured inputs and outputs and ecological services, the satiability of consumer wants, and the ethical questions raised by unequal environmental impacts on different parts of society. These aspects call into question the marginal techniques outlined above, and have led some people to identify a distinctively different, but inherently diverse, specialisation called *ecological economics*. This transdiscipline takes a broader approach to problem solving, often using techniques avoided by most economists, in particular those based on measurements in physical units like land area, mass or energy, rather than money.

3. Concepts, controversies, contributions and perspectives

As a Johnny-come-lately, the economics of the environment has not always received the attention it deserves. Until the early 1970s, major fields of study in economics such as international trade and economic growth completely ignored the environmental dimension. Thirty years ago almost all theoretical models and empirical analyses failed to consider the relationships, connections and causal loops between the environment and the economy. Fortunately, this has changed dramatically. Beginning with the first oil price shocks in 1972–73 that led to a widespread concern over scarcity of non-renewable resources (especially fossil fuels), economists have paid much more attention to resource and environmental issues. More recently, global environmental problems such as stratospheric ozone depletion, climate change and biodiversity loss have forced many economists to face the fact that the economy and the environment interact in important ways.

New ways of thinking about the connections between the environment and economy have also been stimulated by the debate over whether economies are on a sustainable path, and to what extent we can substitute between produced or manufactured capital and outputs, and natural resources and ecological services. This interlinked debate on sustainability and substitutability, and a plethora of studies on the effects of economic growth and the environment and many other economy-environmental issues, have helped make the economics of the environment today a well-established field of study within economics.

Concepts

The economics of the environment incorporates many of the concepts and modes of analysis used in other areas of economics. First and foremost, it starts with the premise of human supremacy, in the sense that almost all analysis is anthropocentric, and seeks ways of improving outcomes for people. This does *not* imply that other organisms are unimportant, but rather that their only economic importance is to generate, directly or indirectly, human values. For example, economic analysis may show that blue whales have a much greater value left unmolested (mainly the 'existence value' that people get from just knowing the largest living animals continue to live), than if they were hunted for their worth as meat and oil products. This comparison, however, would be purely from a human perspective: only values *to people* are counted.

Given the tendency for economists to optimise and analyse, the economics of the environment shares with mainstream economics the almost universal assumption of *rational* behaviour, that is, that people consistently do the best they can ('maximise their utility'), given the constraints they face. From this perspective, human behaviour can be analysed by specifying an objective function or set of goals, and the constraints that decision-makers face in choosing their most desired or optimal decision. When specifying objectives, economists will often assume *non-satiation:* the notion that people always prefer more to less (though as already noted, ecological economics may depart from this). Thus, when specifying that firms maximise their profits, a standard economic assumption is that firms always prefer more profits to less.

Another important set of concepts in the economics of the environment relates to *dynamic analysis*, or the study of how variables and outcomes change over time. This is particularly important in resource economics, where resource stocks typically change over time because of natural growth and harvesting flows. The concepts of stocks and flows also apply to environmental amenities and the ability of the environment to assimilate wastes, and have been used to construct models of the feedbacks between the environment and the economy.

A common feature of dynamic analysis is the *discounting* of benefits and costs, whereby a dollar received in the future is worth less to someone than a dollar received today. People do discount the future, but it can lead to some perverse results. For example, when comparing alternatives where one choice is very high returns today and zero or even negative returns in the future, and another choice is small or moderate returns forever, the former may dominate even at low rates of discount if the assumed objective is to maximise the discounted net returns over time. In other words, suppose the technology existed to triple world net output for the next 200 years by extracting energy from the sun at a low cost, but in 200 years time it would result in the sun exploding (Page 1977: 250). Simply (and inappropriately!) discounting the increased output at anything more than 0.2% per year can be shown to recommend using the technology, because the end of civilisation that the explosion causes is so far in the future (and is thus heavily discounted).

Contributions

The economics of the environment offers a number of important contributions to understanding nature. Foremost is its ability to link human systems to environmental systems via individual and collective incentives. For example, if there are no formal or informal property rights to an environmental asset, so that anyone can use it as she or he sees fit without constraint, then there will arise the so-called 'tragedy of the commons'. This is a form of overuse or even extinction of the resource which may be succinctly described as 'everybody's access, nobody's property'. By understanding and offering solutions to mis-specified property rights and incentives, economics offers a powerful framework for managing natural resources better.

Property rights literature has also been used to develop innovative instruments or mechanisms for pollution control. For example, coal-fired electric utilities in the United States operate under a 'cap and trade' program where they have fixed amounts of sulphur dioxide, in the form of tradeable pollution permits, that they are allowed to emit per year. To help ensure that the overall costs of emission control are minimised, utilities that face a high cost of reducing emissions can buy the right to emit sulphur dioxide from utilities with lower costs. The result is that for a given total emissions level, overall control costs are reduced. This is because high-cost utilities are able to buy emission permits for a price lower than it would have cost them to reduce their own emissions, and low-cost utilities undertake the emission control. Such an approach works well where emissions are uniformly dispersed, so causing identical damage regardless of their geographical origin, and avoiding the creation of localised 'hot spots' of excessive pollution.

The economics of the environment has also provided a framework for improved natural resources management in fisheries, forests, and water use, and in the extraction of non-renewable resources. *Bioeconomic analysis* incorporates biological models of how different living resources grow into economic models. *Dynamic optimisation* — maximising the discounted sum of net benefits from now until some future time horizon — allows resource managers to consider various alternatives and scenarios for the future, and develop operational strategies. Such approaches, coupled with risk assessment, provide a powerful set of tools to improve overall natural resource management. By incorporating the human dimension in resource management, economics provides an understanding of how to address management problems that arise from resource use.

A major contribution towards understanding the environment is the development of detailed models that link economic systems to environmental outcomes (and vice versa). These models come in many different forms, and include studies that relate changes in national income to environmental degradation, studies that evaluate input intensities (ratios of inputs, such as energy use, to gross production or output) and emission intensities (ratios of emission to gross production) over time, and comprehensive models that link sectoral activities in an entire economy to environmental services. These approaches have been used, for example, to develop scenarios that assess the effects of climate change (global warming) on economic activity. Such models help to assess the trade-offs between the present and the future, such as the costs associated with business-as-usual management versus proactive decisions to slow down climate change.

Another important development in the past 30 years in the economics of the environment is *non-market valuation*. It involves several different methods that try to measure marginal external costs, that is, place values or 'prices' on environmental assets. One of the approaches, called *hedonic analysis*, tries to relate decisions people make over purchased goods and services to environmental quality. For instance, an estimated lower limit of the costs of air pollution can be made using economic models that account for differences in prices for similar houses in neighbourhoods with different levels of ambient air quality. In another approach, a lower limit on the value to users of a lake for recreational purposes can be found by asking them the travel costs they incurred in getting to the lake. In this *travel-cost method* the user's value is deemed to be at least as much as his or her costs of getting there.

For environmental assets that are not directly used, *stated preference* techniques (mainly two survey methods called *contingent valuation* and *choice modelling*) have been applied, whereby individuals are given information and asked to respond to questions that elicit their values, such as their willingness to pay to conserve wilderness areas. These approaches are not without their flaws, but have proved very important in helping make decisions about the use (or non-use) of environmental assets.

Controversies

As in any discipline, practitioners have debated the validity and efficacy of various approaches and methods. A key debate in the economics of the environment that goes back to the very beginnings of the subdiscipline is the appropriateness of discounting the future. Some have argued that discounting places a lesser weight on future generations than on the current generation, and is thus 'morally reprehensible'. In response to this criticism, economists have developed approaches whereby the discount rate varies over time, and often undertake analyses with a range of discount rates.

A more recent controversy has revolved around the ability of surveys that elicit stated preferences to provide meaningful values. These critiques and cumulative experience have led researchers to refine the approaches to yield more robust estimates of values, although critics still remain sceptical as to what values are being measured.

A fundamental philosophical debate also exists over the ability of human ingenuity and technical progress to address environmental problems. An extreme view is that provided the incentives exist for developing innovations, technology will be developed to address the full range of environmental challenges. The contrary view, seldom expressed by economists (other than ecological ones), is that nature has fundamental thresholds, and there is no assurance that humanly devised systems can overcome the problems these cause, especially those with long time lags between cause and effect. Most economists take a view in between these perspectives: they acknowledge the uncertainties and difficulties with addressing environmental problems (especially threshold effects) after they have arisen, but also recognise that substitutions and innovations in the past have proved effective at mitigating environmental damage. This debate is also mirrored in controversies over whether economic growth in a country is generally good for the environment. Such controversies have been stimulated by studies claiming to show an 'inverse-U' relationship between environmental degradation and per capita income, where environmental damage first increases and then declines as a country gets richer.

Questions of measurement and environmental performance have arisen regarding the system of national accounts that is the database for most macroeconomic policy. So-called 'green' accounts have been developed that primarily consist of 'satellite' accounts documenting changes in quantities or stocks of important natural resources such as forests or water. In addition, expenditure on the environment, such as pollution control costs, is often separated from traditional accounts to permit further analysis. A much more controversial development has been the development of alternative indicators of overall national welfare, or whether or not national development is sustainable, that include the 'genuine progress indicator' and a measure for 'genuine savings'. Although such measures are now regularly calculated for many countries, they are still viewed as 'sideshows' by many economists in comparison to traditional indicators such as gross domestic product.

Finally, in common with many other fields in microeconomics, the presumption that governments always act to improve social welfare has been comprehensively challenged by many studies of the political economy of environmental policies. These have shown the importance of lobby groups, political self-interest and even outright corruption in explaining which policies are actually adopted.

4. Methods and tools of analysis

The main methods and tools used by the economics of the environment, with examples of their use, are detailed below.

Primary data collection used to be almost exclusively done by various types of *survey instrument,* used either by regular (and often mandatory) government or firm data gathering, or by economists researching a particular issue. Examples of the former would be government data on annual depletion and discoveries of mineral or hydrocarbon reserves, or a firm's reports of their own emissions as part of environmental accounting. An example of the latter would be a contingent valuation survey designed specifically to enable use and non-use values of some threatened wetlands to be measured in monetary units. An example of both would be asking firms how much they spend on pollution control, a question which might be posed either regularly by a government agency, or as part of a one-off research project. However, in the last decade or so there has been an increasing amount of primary data gathered by *laboratory experiments,* where people make behavioural choices, in return for modest amounts of money, in a controlled laboratory situation that mimics a market transaction of interest. An example would be an experiment that might be used to test how people would bid in various types of auctions of tradeable pollution rights.

Secondary data collection is mainly by electronic transfer or physical copying and entering of existing survey data held by government and firm sources. It is the only form of data collection that many mainstream empirical economists ever do, though by comparison, environmental economists need to collect primary data more often. Examples of secondary data would be consulting annual economic reports of national statistical agencies in a library, and downloading a spreadsheet of data on many countries' economic growth and environmental performances from a World Bank website.

Statistical analysis of primary and secondary data sources is the mainstay of much empirical analysis in the economics of environment. Like mainstream economists, environmental economists need to use special statistical techniques called *econometrics* to deal with the many economic data series that show rising or falling trends over time, without a clear direction or even existence of any causal connection among them. For example, population, GDP per capita and the number of cars per household are all typically rising, and econometric analysis is needed to assess the separate contributions of each factor to rising air pollution in cities. Another example would be to analyse the effects of income per capita and various social factors on measures of overall environmental quality, to see if any inverse-U relationships exist between income and environmental degradation.

Mathematical modelling, with emphasis on dynamics, is a key technique for trying to understand the inner workings of economy–environment interactions. Often statistical analysis will show that parameter A influences factor B, but only mathematical modelling can explain why it does so. Ideally models would use parameter values obtained from primary or secondary sources to produce an empirical result about the real world, but much mathematical modelling in economics is purely theoretical with no empirical testing, and environmental economics is no exception. Examples of mathematical models abound both for microeconomic questions, such as the optimal time path for planting and felling a forest, or for extracting ore from a mine; and for macroeconomic questions, such as recalculating national income measures to include the effect of environmental degradation and resource depletion.

Lastly, *cost-benefit analysis* (CBA) is a hybrid of the above techniques, designed to quantify in money units the net benefit or cost of a particular project, such as a new environmental regulation or a new emission control investment. A key aspect of an environmental CBA is the choice of non-market values to use, and the decision of how many second- and third-order effects to include in the analysis. For example, in evaluating the non-market effects of a road construction project, does one include just the reduction in travelling time of existing drivers, or the compensating increase in congestion resulting from new drivers taking to the roads once travel is faster?

Acknowledgements

We are grateful for the helpful comments and suggestions of Robert Costanza, Richard Damania and Jo McAllister.

5. Further reading

Blaug's (1997) history of economic theory succinctly explains the main contributions of Smith and other great economists. Tietenberg (2002) is an introductory textbook in environmental economics that has a wealth of empirical and institutional detail, especially concerning the USA. We recommend van den Bergh (1996) as a good introduction to ecological economics, and Garrod and Willis (1999) as an excellent first guide to environmental valuation.

5 THE ENVIRONMENT AND HUMAN HEALTH

A.J. McMichael

1. Defining 'environmental health'

The environment, considered in relation to health and disease, evokes images of polluted air, chemical residues in food, radioactive wastes and so on. Conventionally, 'environmental health' refers to research and policy in relation to health risks posed by ambient physical, chemical or microbiological agents. These exposures usually impinge on all members of a group, community or population. That is, environmental health hazards are rarely under personal control, unlike behavioural risk factors such as smoking, alcohol consumption and unsafe sex.

This chapter concentrates on the aspects defined above, but we should note several other facets of this broad topic. First, the social, physical and aesthetic environment can also influence sense of well-being and mental health. (The condition 'seasonal affective disorders', or SAD, is well-documented in the cold dark winter months of northern Europe). Indeed, culture and associated social norms are an important, pervasive, influence on consumer choices and behaviour. Consumerism, economic priorities and technological choices also directly contribute to the pattern of local and, increasingly, global environmental exposures. Second, and relatedly, the environment can be a force for good health (a 'melior') as well as for poor health (a 'hazard'). (In cities, for example, bicycle paths encourage people to cycle to work — as happens, par excellence, in the Netherlands. This is beneficial to physical health, relative weight, and, probably, mental health). Third, environmental scientists and ecologists also use the phrase 'environmental health' to refer to the health of the environment. Further, over the past decade there have been attempts to describe the nexus between environmental health and human health in terms of 'ecosystem health'. (Indeed, a new international journal *EcoHealth*, launched in 2004, deliberately seeks to achieve this type of synthesis.)

Not all environmental health hazards are human-made. Indeed, pre-industrial concerns focused more on hazards from the natural environment. Over two thousand years ago, in classical Greece, Hippocrates wrote in *Airs, waters, places*:

> Consider the seasons of the year, and what effects each of them produces ... then the winds, the hot and the cold, especially such as are common to all countries, and then such as are peculiar to each locality ... concerning the waters which people use, whether they be marshy and soft, or hard and running from elevated and rocky situations, and then if saltish and unfit for cooking.

Natural environmental influences on health include extremes of weather, locally circulating infectious agents, physical disasters, and local micronutrient deficiencies in soil. For example, almost one-fifth of the world's population lives on ancient, leached, and often mountainous soils that are deficient in iodine. This puts many such populations at risk of dietary iodine deficiency disorders, including goitre, reproductive impairment, intellectual impairment and neuromuscular disorders (Hetzel and Pandav 1994).

A broader view of environmental health encompasses urban design, housing quality, material circumstances and general socioeconomic status. Further, the recent emergence of global environmental changes (such as human-induced global climate change), as a new and unfamiliar category of environmental risk to human wellbeing, health and survival, extends the topic of environmental health even further — and thereby connects it more directly with the discourse on sustainable development.

An even more radical view of environment is that it comprises all non-genetic factors that influence health. Cigarette smoking, oral contraceptive usage and so on thus become yet another part of an individual's environment. Conventionally, however, the boundary of the 'environment and health' domain is set so as to distinguish it from both personal-behavioural exposures (or risk factors) and the health influences of the social environment, acting via either materialist or psychosocial pathways. We are thus concerned primarily to understand and ameliorate the risks to health posed by ambient, pervasive, environmental hazards.

The principal research disciplines that inform this topic are (i) epidemiology, which describes and quantifies risks to health; (ii) toxicology, which elucidates the metabolic derangement and resultant biological damage due to exposure to a noxious chemical; (iii) microbiology, which determines the identity and biological characteristics of infectious agents; (iv) environmental sciences, which describe the processes by which environments vary naturally or are perturbed by human intervention, and the consequent interactions between environmental media and systems, and (vi) the social sciences that enrich our understanding via historical and sociological perspectives.

Epidemiological research provides the bedrock information, necessary for the identification, quantification and practical amelioration of environmental risks to health. This chapter will give particular attention to this scientific discipline. Epidemiology is defined by the '3 Ds': it studies the distribution and determinants of disease in human populations.

Epidemiological studies of the health consequences of environmental exposures are carried out at several levels. Descriptive studies compare environmentally-related rates of disease or premature death – such as comparing the prevalence of dental caries in communities with different levels of fluoride in their drinking water. At higher resolution, information can be collected at local or individual level, and then used to estimate dose-response relationships – such as follow-up studies of intellectual development in childhood in relation to antecedent measures of individual blood lead concentrations. Some epidemiological studies are enriched by collaboration with biomedical research disciplines, to elucidate the biological pathways that mediate health-damaging exposures. Meanwhile, as new attention focuses on estimating the likely future health risks of global environmental changes such as climate change, so the application of mathematical modelling becomes more prominent.

2. History of environmental health: ideas, practice, research methods

Concern over environmental conditions has substantially shaped the historical evolution of public health and its core research discipline, epidemiology. In 19th-century Britain, there was a need to understand and remedy, for example, toxic hazards in locally-brewed alcoholic drinks, the cholera outbreaks in mid-century London, differentials in death rates between different residential and socioeconomic groups, and various specific occupational exposure hazards.

During earlier centuries, longstanding ideas persisted about disease reflecting God's retributory judgement on the human condition. However, in the 17th century new philosophical perspectives emerged, as the foundations of modern empirical science were laid in western Europe. Francis Bacon argued for scientific enquiry based on empirical observation and comparison. Rene Descartes propounded a reductionist framework for studying the external 'machine-like' world.

This Enlightenment thought, the rise of inductive logic (drawing generalised inferences about relationships and physical laws from specific sets of observations), and a utilitarian approach to the fruits of scientific enquiry nurtured a more activist management of the environment. Consequently, the social hygiene movement arose in late 17th-century Europe, seeking to improve and cleanse the environment. Major social expenditure was required, for works ranging from draining marshes to removing urban refuse or improving roadways.

After the French Revolution, more egalitarian social ideologies developed. Adverse environmental impacts on the health of urban-industrial populations were recognised. In Britain, the 'sanitary idea' emerged in the 1840s and became linked with ideas of urban sustainability – including ideas of recycling sewage, attaining local self-sufficiency in food production, and achieving full employment. The idea of garden cities became popular, to counter the health-endangering miasmas presumed to arise from dank and dirty urban environments.

Meanwhile, improvements in nutrition, the urban environment and general social progress were ushering in the first public health revolution. Other health gains arose via changes in human social organisation and economic practices. For example, the mechanisation of European agriculture from mid-18th century, with increased fodder production, stimulated a growth in cattle herds. This caused a reduction in human malaria, since the malaria-transmitting anopheline mosquitoes much prefer their blood-meals from cattle than from humans.

The spectacular rise of bacteriology in the 1880s caused a frameshift in perspective. Microbes were deemed to be the primary cause of disease. This potent 'germ theory', along with new theories of cell biology and heredity, awareness of micronutrient deficiencies, and the medicalisation of child-bearing and child-rearing, all refocused the health sciences on the individual. Ideas of shared environmental exposures and their health risks receded. While the germ theory prevailed early in the 20th century, some counterbalancing of perspective arose from new knowledge about the vector-based transmission (by mosquitoes, ticks, biting flies, etc.) of major diseases such as malaria, schistosomiasis, dengue fever, yellow fever and leishmaniasis, leading to new environmental management strategies in tropical and subtropical regions. This included the spraying of mosquito breeding sites with DDT, control of surface water, and the control of alternative mammalian host species.

By the mid-20th century, the spread of industry and motorised transport had greatly increased environmental exposures. Major urban air pollution episodes occurred, and these helped stimulate new environmental legislation in western countries during the 1960s and beyond. Meanwhile, in *Silent spring* (1962), Rachel Carson focused on a new concern — pervasive ecological damage by bioaccumulating pesticides. Humans, she argued, were on notice that such chemicals, rippling throughout nature, would eventually impinge on human biology.

In the 1970s, the growing recognition of acid rain heralded a new awareness that environmental health hazards were transcending landscapes and national boundaries. New concerns also arose about the health risks from cumulative exposures to various types of environmental contaminants, especially the persistent chlorinated hydrocarbons and several of the heavy metals — exposures that can impair the immune, reproductive and neurological systems.

Finally, in recent years there has been rising interest in understanding how human-induced global environmental changes such as climate change, biodiversity loss, and degradation of food-producing ecosystems, affect the risks to, and prospects, for human health. This, more than any of the preceding topics, has underscored the need for a more integrative and interdisciplinary approach in research and in policy development. These emerging, larger-scale, hazards to health arise from the disruption or depletion of complex biophysical and ecological systems, and their elucidation requires more than the single discipline and reductionist approaches that have generally served epidemiologists, toxicologists and environmental scientists well in the past.

3. Environmental health research: some specific issues

Most environmental epidemiological research is necessarily observational — that is, non-experimental. (It is usually neither ethical nor feasible to do an experiment to assess the impact on human health of some putative hazardous environmental exposure.)

The key concepts and postulates of mainstream environmental health research include the following:

- The causal chain extends from drivers (such as pesticide manufacture or fossil fuel combustion), through resultant pressures on the environment, to alterations in composition (or 'state'), and thereby to changes in levels of exposures impinging on humans, resulting in changes in states of health.
- Environmental exposures occur in a variety of configurations – acute, intermittent, recurring and chronic. This exposure pattern has consequences for risks and types of health consequence.
- The effect of the received level of exposure to physico-chemical agents (that is, the 'dose') and the probability, or extent, of health outcome can be estimated as a dose-response relationship.

- Microbiological contamination of air, water and also food displays a general dose-response tendency – although the notion of critical (threshold) dose is often important.
- Environmental exposures often do not act in isolation or separately. Often the net impact on human health is a consequence of biological interaction (synergy or antagonism) between multiple concurrent exposures.

Spatial analysis

Many of the relationships investigated in environmental epidemiology require spatial analysis (Elliott et al. 2000). Simple techniques entail, for example, the use of concentric circles around point sources of environmental pollutants or of residential distance from main highway. As the amount and quality of information on the spatial distribution of the exposure hazard, and of its sources, increases, along with the equivalent information on the distribution of health outcome measures, so the possibilities increase for formal spatial analysis. Geographic information systems (GIS) provide an analytical technique for identifying spatial relationships by integrating 'layers' of spatial data about one or more exposures, other confounding or modifying factors, and health outcomes. Research strategies have evolved over the past decade for the study of small-area environmental epidemiological statistical analysis. Improved techniques for smoothing data and for dealing with auto-correlation and outlier observations are available. The use of such techniques has been aided by the burgeoning computing power available.

4. Urban-industrial environmental hazards

The advent of legislative environmentalism in Western societies during the last third of the 20th century reduced much environmental pollution. Similarly, acid rain has been curbed, the use and release of heavy metals has decreased, pesticide use has been constrained, and the nuclear power industry has become better controlled. These environmental issues predominated at the 1972 UN Conference on the Human Environment in Stockholm. Meanwhile, however, urban transport density and the associated emissions of some exhaust gases have increased. The resultant rises in concentration of photochemical-oxidant air pollutants (especially ozone) and very fine particulates pose particular risks to cardiovascular and respiratory health.

In low-income countries the profile of environmental exposures reflects both the old and the new. Microbiological hazards, especially in drinking water, remain widespread in rural populations and in urban slums. Approximately 40% of the world's population still lacks safe drinking water and 60% lacks sanitation. Domestic air pollution levels are often high, especially where biomass fuels or coal are used for heating and cooking, resulting in high rates of infant and child mortality from respiratory infections. Meanwhile, as industries proliferate, as chemical-intensive agriculture spreads, and as cities fill with cars, trucks and buses, so the ambient environment acquires a range of additional physical, airborne, waterborne and foodborne hazardous exposures. The environmental pressures due to population growth and economic intensification can also exceed critical thresholds so that they then adversely affect physico-geochemical cycles. An important example is the now-widespread public health problem of arsenic-contaminated groundwater from deep tubewells sunk throughout Bangladesh and West Bengal in response to the increasing demand for fresh water (see chapter 8), compounded by increasing levels of faecal and other contamination of surface waters in an impoverished and over-crowded environment.

Today, almost half of all people live in urban environments. The modern urban environment comprises industrial activities, concentrated transport systems, intensive waste generation, and more fluid, non-traditional patterns of social relations and interactions. Some of the resultant risks to health may be overt, as with road trauma or the increase in asthma hospitalisations during air pollution crises. Others, however, are non-acute and more insidious — well illustrated by the deleterious impact of lead exposure upon early childhood intellectual development (figure 5.1; Tong and McMichael 1999).

Housing quality is another important aspect of the urban environment that influences seasonal and other patterns of morbidity and mortality. For example, housing quality, including dampness, may contribute to early-life exposure to fungal spores and to house-dust mites, initiating asthmatic predisposition in children by affecting the early-life maturation trajectory of the immune system.

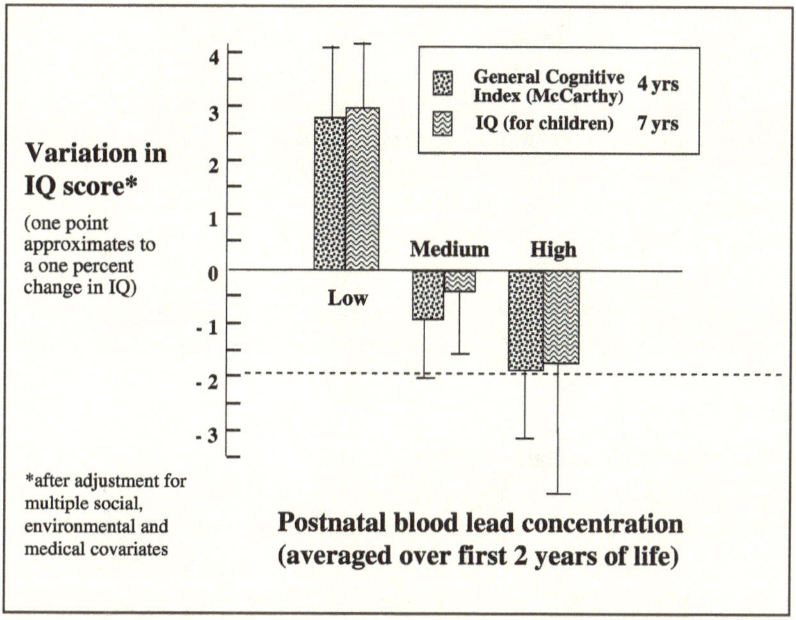

Figure 5.1. Environmental lead exposure and early childhood cognitive development: variations in mean scores (and standard deviation) by level of early-life exposure to lead (from Tong and McMichael 2001).

Two other major examples of today's urban environmental hazards are urban air pollution, and the vulnerability of inner-urban populations to heatwaves.

Urban air pollution

Urban air pollution has, in recent decades, become a worldwide public health problem. Overall, industrial/domestic air pollution from coal-burning has been increasingly replaced by pollutants from motorised transport. These form photochemical smog in summer, and episodes of heavy haze of particulates and nitrogen oxides in winter.

In China, urban air pollution with particulates and sulphur dioxide is increasing sharply, particularly from the industrial use of high-sulphur coal. However, emissions from domestic cooking and heating fuels are also increasing. The World Bank has recently assessed that morbidity and mortality in Chinese cities will increase steeply, because of air pollution exposures, over the coming two decades, with a tripling of chronic bronchitis cases and annual bouts of respiratory symptoms (World Bank 1997).

Epidemiologists have developed a diverse and sophisticated set of methods for assessing the health impacts of air pollution. Nevertheless, difficulties persist in exposure assessment, in differentiating between the effects of acute and chronic exposures, in elucidating interactive effects between coexistent air pollutants, and because the profile of air pollution keeps changing. That is, today's research results often refer to yesterday's pollution levels.

Thermal stress: urban vulnerability and mortality

Severe heatwaves adversely affect health – particularly in the centre of large cities, where temperatures may be higher than in the suburbs and the surrounding countryside, and where the relief of night-time cooling may be lessened. This 'heat island' effect reflects the heat-retaining concrete, asphalt and masonry structures of most inner cities, the physical obstruction of cooling breezes and the lack of parks and trees.

In July 1995, over 460 extra deaths were certified as due to the effects of an extreme heatwave in Chicago, when temperatures reached 40°C. Studies of such episodes elsewhere in North America and Europe have shown that those most vulnerable to heat-related illness and death are the elderly, the sick and the urban poor. These studies underscore the mix of determinants of mortality excess during heatwaves: the urban environment itself as part of modern human ecology, the cultural and technological adaptation to coping with heat, and the characteristics of individuals that increase vulnerability.

5. 'New and resurging' infectious diseases: a global phenomenon

Infectious diseases receded in Western countries throughout the latter 19th and most of the 20th centuries. This was, initially, mostly because of urban sanitation, improved housing design, personal hygiene, antisepsis in clinical medicine, and vaccination. The antibiotic era consolidated these gains. However, the receding tide apparently turned in the late 20th century, and there has been much recent talk of 'new and resurgent' infectious diseases.

Populations everywhere are becoming interconnected economically, culturally and physically, enhancing the mixing of people, animals and microbes from all geographical areas. Human mobility has escalated, in volume and speed, between and within countries. Long-distance trade facilitates the geographical redistribution of pests and pathogens. This has been well illustrated in recent years by the HIV pandemic, the worldwide dispersal of rodent-borne hantaviruses, and the rapid dissemination of a new epidemic strain of bacterial meningitis along routes of travel and trade. An important corollary of this is that unexpected outbreaks of infectious diseases, such as SARS (severe acute respiratory syndrome), can now have a paralysing effect on tourism and trade, with potentially disastrous economic consequences.

Meanwhile, the relaxation of traditional cultural norms is yielding newer, freer patterns of human behaviour, including sexual activities and illicit drug use. Modern medical manoeuvres create new ecological opportunities for viruses and prions; hospitalisation practices have done similarly for various bacteria. Over the past half-century antibiotics have been used widely and often unwisely (including for livestock and agricultural purposes), thus fostering drug-resistant organisms.

The occurrence of infectious diseases is widely influenced by land clearance in populous regions of the developing world and by the extension of irrigation. In Sudan, for example, schistosomiasis appeared within several years of the start of the large irrigated Gezira cotton project. Various viral haemorrhagic fevers have emerged over recent decades as intensified land clearance and habitat disruption, especially in South America, have exposed human populations to new viruses that previously circulated exclusively within wilderness ecosystems. For example, the Junin virus, which causes Argentine haemorrhagic fever, naturally infects wild rodents. However, the extensive conversion of grassland to maize cultivation in recent decades, stimulating a proliferation of the virus-bearing mice, has exposed farm workers to this 'new' virus.

In summary, the globalisation of our economic activities and culture, worldwide urbanisation, increasingly rapid and long-distance mobility, growth in refugee movements, and the intensification of food-producing systems, are creating a world of increasing opportunity for established and potential human pathogens.

6. Environmental health research: the wider dimensions

We live at a critical historical time, in which human domination of this planet now poses serious problems for future generations (Vitousek et al. 1997). Today's world is undergoing a rapid increase in connectedness, in the economic, political, cultural, physical and electronic realms. The resultant changes in economic structures, production, trade and financial investment have various important consequences for the environments within which human populations live, and therefore for health (Beaglehole 2003).

Before discussing the health risks from this unprecedented planetary-scale disruption of various of Earth's natural systems, let us briefly note the complex relationships between environmental degradation, population size, poverty and health. This is a controversial topic, and it reminds us that we will only solve many of the world's environmental problems within a framework of poverty alleviation and equity.

Some of the relationships between environmental conditions, socioeconomic circumstances, demographic change and human health are immediate: for example, poverty today causes malnutrition today. Other relationships involve long time lags: for example, current poverty contributes to the need to clear local forests for fuel and to farm marginal lands, leading to ecological attrition and hunger in future. Overall, there is no simple linear causal chain connecting these variables. Population pressure and poverty among rural populations often lead to land degradation and deforestation, and, within the poorer sections of the urban environment, they predispose to other adverse environmental exposures (physical hazards, high exposures to polluted air and unsafe drinking water). Poverty influences fertility rates, and vice versa. Environmental degradation often causes further impoverishment — for example, by reducing food yields, depleting fuelwood, or potentiating destructive floods — and it can also impair health via increased exposure to vector-borne infectious disease, nutritional deficiencies, and toxic environmental pollutants.

One important consequence of the rapid expansion and intensification of the human enterprise — a function of population size and economic practices — is that new pressures are being put upon various aspects of the world's environment. The spread of large-scale agribusiness and monoculture crop production is exacerbating soil erosion and land degradation. Freer access to the world's forests by logging companies is creating problems of erosion, flooding, silting of reservoirs. The continued building of large dams not only displaces many millions of persons, but creates breeding sites for various infectious disease vectors, and carries the risk of occasional massive dam-break disasters. The expansion of long-distance trade creates new opportunities for the spread of pests and pathogens, as evidenced by the introduction of the mosquito species *Aedes albopictus* via cargoes of used tyres, from Asia to South America, the USA and parts of Africa.

More fundamentally, we have apparently begun to exceed the capacity of the biosphere's sources and sinks — that is, the life-support systems upon which human wellbeing and good health depend in the long term. Around the world, many other river systems are shared uneasily between neighbours in unstable regions, including the Ganges, Mekong, Jordan, and Tigris and Euphrates rivers (Gleick 1998). Several ambitious global assessments have estimated that, globally, humankind is now living in a situation of significant and increasing 'ecological deficit', as natural environmental stocks decline (Wackernagel et al. 2002).

7. Global environmental change and health

The two best-defined global environmental changes, the depletion of stratospheric ozone by the emission of ozone-destroying gaseous emissions (especially chlorofluorocarbons, CFCs) and the accumulation of heat-trapping greenhouse gases (GHG) in the lower atmosphere, each entail changes in 'global commons'. That is, although the gaseous emissions arise from diverse localised sources, in all continents, their environmental impact is of a diffuse globalised kind. These local emissions thus result in integrated global changes — and these pose hazards to human health.

Global climate change

Climate scientists forecast that the continued accumulation of heat-trapping greenhouse gases in the troposphere will change global patterns of temperature, precipitation and climatic variability over the coming decades (McMichael et al. 2003). A rise of 1–3°C over the coming half-century, greater at high than at low latitudes, would occur faster than any rise encountered by humankind since the inception of agriculture around ten thousand years ago. (While the protracted warming after the last glaciation proceeded slowly, occurring over several thousand years, there were apparently acute episodes when global temperatures swung by several degrees in less than a century.)

The UN's Intergovernmental Panel on Climate Change and various other national scientific panels have assessed the potential health consequences of climate change (McMichael et al. 2001, 2003). These risks to human health will arise from increased exposures to thermal extremes and from regionally variable increases in weather disasters. Other risks would arise from the disruption of complex ecological systems that determine the geography of vector-borne infections (such as malaria, dengue fever and leishmaniasis), and the range, seasonality and incidence of various foodborne and waterborne infections, the yields of agricultural crops, the range of plant and livestock pests and pathogens, the salinisation of coastal lands and freshwater supplies due to sea-level rise, and the climatically-related production of photochemical air pollutants, spores and pollens.

Model Estimate: Current Dengue Risk Region, Australia

Scenario: medium GHG emissions

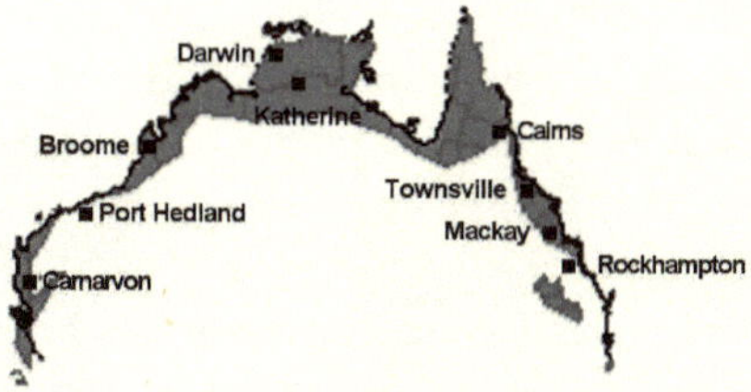

Scenario: high GHG emissions

Figure 5.2. Dengue fever: Estimated changes in geographic regions climatically suitable for maintenance of the vector mosquito, *Aedes aegypti*, 2050 (high and medium emissions scenarios), relative to present climatic suitability (top map).

The wider public and decision-makers are asking scientists for estimates of the likely population health consequences of these great and unfamiliar changes in today's world. Public health scientists now face the task of estimating, via interdisciplinary collaborations, the future health impacts of these projected scenarios of climatic-environmental conditions. Mathematical models have recently been used, for example, to estimate how climatic changes would affect the potential geographic range of vector-borne infectious diseases. Figure 5.2 illustrates some recent modelling of future climate change impacts on the potential transmission of mosquito-borne dengue fever in Australia.

Illustrative of this increased public awareness is the fact that the World Health Organization's estimation of the 'global burden of disease' for the year 2000 included an estimation of the burden attributable to climate change scenarios currently and over the coming decades. Thus, WHO's *World Health Report 2000* estimates that, by 2000, climate change was already responsible for 2.4% of diarrhoeal disease worldwide and up to 6–7% of malaria and dengue in specified groups of countries. Meanwhile, recent evidence suggests there have been climate-attributable trends in tick-borne encephalitis in Sweden (Lindgren & Gustafson 2001), cholera in Bangladesh (Pasqual et al. 2000) and, more contentious, malaria in eastern Africa (Patz et al. 2002).

Stratospheric ozone depletion

Depletion of stratospheric ozone by human-made gases such as chlorofluorocarbons has occurred over recent decades and is likely to peak in about 2020. Ambient ground-level ultraviolet irradiation is estimated to have increased consequently by 5–10% at mid-to-high latitudes over the past two decades. Scenario-based modelling, integrating the processes of emissions accrual, ozone destruction, ultraviolet radiation (UVR) flux, and cancer induction, indicates that European and US populations will, under the severally-amended Montreal Protocol, experience a 5–10% excess in skin cancer incidence during the middle decades of the coming century (Slaper et al. 1996).

Less certainly, greater exposure to UVR will increase the occurrence of various disorders of the eye, particularly cataract of the lens. There are also questions about the consequences of UVR-induced suppression of immune system activity – although some consequences could be beneficial to health (such as a reduced level of auto–immune diseases).

Biodiversity losses

As human demand for space, materials and food increases, so populations and species of plants and animals are being rapidly extinguished. An important consequence for humans is the disruption of ecosystems that provide nature's goods and services. Biodiversity loss also means the loss before discovery of many of nature's chemicals and genes, of the kind that have already conferred enormous medical and health benefits.

Meanwhile, invasive species are spreading worldwide into new non-natural environments via intensified human food production, commerce and mobility. The resultant changes in regional species composition have myriad consequences for human health. For example, in eastern Africa's Lake Victoria the spread of the water hyacinth, after its introduction from Brazil as a decorative plant, now provides a breeding ground for the water snail that transmits schistosomiasis and for the proliferation of diarrhoeal disease organisms.

Finally, humankind is rapidly changing the cycling of major elements within the biosphere. Since the 1940s there has been a sixfold increase in the global annual human fixation of nitrogen (i.e. its conversion from inert to biologically active form). This amount, mostly from the use of nitrogenous fertilisers, now exceeds the annual natural yield of fixed nitrogen. Likewise, we have markedly altered the geochemical cycling of sulphur. By affecting the acidity and nutrient balances of the world's soils and waterways, these two chemical changes are likely to impair the capacity of terrestrial and aquatic food-producing systems.

8. Conclusion

A fundamental point of this chapter is to make clear that the environment impinges on human wellbeing and health via diverse pathways. The environment itself is a multi-faceted entity, comprising physical, chemical, biological, social and aesthetic aspects. Environmental influences on human health are both positive and negative.

Much of the conventional focus of concern has been, and is, on the toxicological hazards posed by chemical and physical pollutants of air, water, soil and food. However, the environment also includes a variety of life-supporting systems and processes that are of fundamental importance to human survival and to the achievement and sustaining of good health. The recent recognition of the ways in which changes in human ecology (such as social diets and physical activity patterns) can affect health outcomes has underscored this need to think more ecologically, and to use more integrated, holistic, approaches to research and policy development.

The advent of unprecedented global environmental changes, such as global climate change, has reinforced this need to think in larger-scale, trans-boundary, terms. Hence the recent rise in research interest in understanding the health consequences of changes in patterns of international trade, travel, the movement of plant, animal and microbial species, disruptions to ecosystems, and changes to atmospheric composition. To an increasing extent the environmental determinants of population health arise from beyond the physical boundaries of the particular affected population.

9. Further reading

The article by Ahern and McMichael (2002) gives a good overview of the contemporary issues and relevant research methods in environmental health. *Environmental Health* (an open access journal on the web at http://www.ehjournal.net/home/) contains a good international spread of peer-reviewed papers on environmental medicine, epidemiology and toxicology. No password or user fee is needed. Pitched at an international readership, *Basic environmental health* by Yassi, Kjellstrom et al. (2001) offers a clear and accessible introduction to the topic, with diverse examples from around the world. For an overview chapter on environment and health in one of the world's two leading public health textbooks, see McMichael (2002). The paper by McMichael et al. (2001) provides an up-to-date review of the world literature on current and future health impacts of global climate change.

6 POLICY ANALYSIS, ENVIRONMENT AND SUSTAINABILITY

Stephen Dovers

The formulation of policies and their implementation through policy and institutional systems is critical to the collective achievement of any social goal, including that of protecting the environment. In achieving positive change in the environment, and more importantly human interactions with natural systems, the usefulness of knowledge from formal disciplines or other knowledge systems will be determined by its incorporation in policy systems designed to change the behaviour of individuals, households, firms, communities and organisations within society.

This chapter introduces definitions and frameworks for understanding policy and policy analysis. Consistent with shifts in the way environmental policy is understood, section 2 notes major features of the broader, integrated policy agenda of sustainability (or sustainable development). 'Sustainability' can be defined as a system property or a long-term social goal, and 'sustainable development' a policy agenda and series of initiatives aimed at enhancing that property and/or addressing that goal. Sustainability is a higher order social goal, on par with justice, peace, democracy or the rule of law, and therefore not to be expected to be precisely defined or achieved quickly. The idea of sustainability has deep and diverse intellectual roots, but was only articulated as a goal of international and national policy in the 1990s. Section 3 illustrates the complexity of policy analysis and of sustainability policy by exploring, in more detail, policy instrument choice and public participation.

The chapter is largely about *public policy*, or policy processes, decisions and implementation by governments, even though these activities are often undertaken in partnership with the private and community sectors. This focus limits the size and detail of the chapter, but also mirrors the focus of the discipline of public policy – environment and sustainability are areas where governments must play the central coordinating role.

1. Defining 'policy'

Considine (1994: 2) observed that policy 'is a deceptively simple term which conceals some very complex activities'. The term is often used loosely and in ambiguous ways. The public policy literature goes into more detail, but does not provide tight definitions. Davis et al (1993: 15) define public policy as *'the interaction of values, interests and resources, guided through institutions and mediated through politics'*, adding that policy is never one decision or action, that many actors are involved, and that it is best viewed as a process over time. Often, policy is discussed in simplistic terms, without being precise about specific parts of these complex processes. The following terms may clarify what considering 'policy' can involve:

A *policy* is an avowal of intent, a recognition of a problem and a statement of what might be done about it, generally but not always issued by government. A policy *program* refers to a set of activities being implemented together.

Management refers to on-ground actions taken in accord with, or sometimes at odds with, a particular policy, often implementing a policy instrument.

Policy instruments are the specific measures used to give effect to a policy (e.g. a regulation, tax, or education campaign).

A *policy system* is the complex interaction of multiple elements of governmental and related structures and information flows, out of which policies emerge.

Policy processes are more specific and variable arrangements for a particular policy within a policy system concern.

Institutions are durable rules and patterns of behaviour determining interactions and transactions (material, economic, informational) within society, whereas *organisations* are the more obvious and more changeable manifestations of these institutions (agencies, NGOs, etc). Institutions, organisations and policy are related phenomena manifesting different aspects of complex institutional systems

A *policy community* comprises the individuals and groups (*policy actors*) inside and outside of government engaged in discussion around a policy, whereas *policy network* refers to the smaller group that formulates and decides policy. An epistemic community is a knowledge-based group involved in policy processes (e.g. ecologists in biodiversity debates).

When considering policy, the *locus of responsibility* for decisions and actions is critical — while a policy network may formulate policy and many agents may implement it, a particular agency or individual normally has the legal responsibility for decisions (e.g. a government minister with the power of final approval for a land clearance permit). The legal basis of policy actions is also important: in the environmental domain, *statute law* is usually the source of legal mandate, although common or customary law may be relevant. *Jurisdiction* is important in this sense, as policy is closely tied to the political and legal basis of power: international, regional, national, state/province or local.

The important parts of a policy and institutional system will vary from country to country, and a good understanding of the specific political, administrative and legal features of the relevant jurisdiction is necessary to understand or engage with the policy system. For example, countries with a central government (e.g. New Zealand, United Kingdom) have some important differences compared to federal systems (e.g. Malaysia, Australia, Canada) where powers in environmental policy are shared between national and state/provincial governments. Similarly, the importance of government and the courts in environmental policy varies across political systems.

Differences in meaning of key concepts across contexts must also be accounted for when comparing systems. This may include the relative influence of interest groups, timing in the political cycle, political and climatic events, legal traditions, informal customs of politics and policy-making. For sustainability policy, the biophysical and ecological context, which often defines the particular issues that are most prominent (air pollution, biodiversity, soil erosion, etc.) is also important. Such differences matter especially when considering policy or analysing it across jurisdictions.

The roots of policy analysis

Modern policy analysis arose after World War II, as social scientists tried to or were called upon to contribute to expanding government policy-making and intervention in society by applying more rigour ('craft') to the 'art' of politics and government (e.g. Lasswell 1951). The field is variously called policy analysis, policy sciences and public policy, with different intentions and directions evident. Over the years, the field has expanded and comprises many sub-sets and approaches (e.g. Daneke 1989). To explore this we can consider four questions: Who does policy research? What is its intent? What do they do? And what 'enduring questions' persist in theory and practice?

First, who does policy research? Brunner (1991: 65) identifies five strands:

- public affairs (philosophers)
- policy analysis (economists)
- management science (public and business administrators)
- policy studies (political scientists)
- socioeconomics (sociologists)

There are other players as well — for example, lawyers, institutional theorists, and in particular public servants who may have no clear disciplinary base but certainly 'do' policy analysis. 'Public policy' is itself a recognisable field, but draws on many disciplines, and is in fact an interdisciplinary field rather than a clear discipline. A policy orientation can in fact serve as a mechanism for interdisciplinary focus on sustainability (Dovers 2003). The many disciplines and groups involved in policy analysis organise and communicate through professional and academic societies, government agencies, non-government organisations, and the private sector.

The intent of policy analysis varies, depending on aim, methods and affiliation of the analyst. One split is between *description* and *analysis*, and across these whether *prescription* is attempted. Hogwood and Gunn (1984: 29) identify two categories that overlap in the area of evaluation:

Policy studies:

- study of policy content
- study of policy process
- study of policy outputs
- evaluation

Policy analysis:

- evaluation
- information for policy-making
- process advocacy
- policy advocacy, with either the analyst as political actor or the political actor as analyst.

One important difference here is between analysis that recommends particular policy *instruments*, and that which recommends *processes*. The role of analysts and whether they are independent, or aligned to government or another policy actor, is also significant. Various approaches are employed in policy research, each with different theoretical constructions. Howlett and Ramesh (2003) separate theoretical approaches according to the scale of analysis (individual, group, institution) and whether analysis is deductive or inductive. Deductive approaches apply generalised or universal assumptions to specific cases, whereas inductive approaches use specific cases to establish generalisable propositions, while being wary of proposing general theories across different contexts. For example, among approaches to policy informed largely by economics, public choice applies deductive theory to the individual scale, while welfare economics constructs theory inductively from that scale. Some common approaches are:

Political science approaches, focusing on the interplay of policy actors, communities and networks. While essential to understanding the political context of policy-making, this may be rather retrospective, contained within notable case study conflicts, and have little prescriptive or predictive value.

Psychological/sociological approaches, examining individual and group processes in a policy context. These are particularly relevant to, for example, consumption-oriented policy, or community-based program design, where people's perceptions play an important role.

Policy/program evaluation, where implementation is assessed for efficacy, outputs, costs, etc, by public agencies or external consultants. Although aimed at administrative efficiency and accountability, given that there are a sufficient number of programs of some similarity in style and goal, this can inform more general understanding.

Public choice, an economic approach based on applying the idea of rational individuals who operate in self-interest, seeking to maximise their utility through making choices in properly functioning markets within constraint sets.

Legal research, varying across 'black-letter law' (i.e. what is written in law) approaches to regulation, administrative law, etc., more 'institutional' approaches looking at the law's role in establishing and maintaining institutions and processes and the codification of principles, and approaches that focus on implementation ('law-in-context').

Institutional approaches, concerned with the way in which deeper institutional arrangements affect the shaping of policy systems and formulation and implementation of policy, and how policy experiences feed back into institutional change (Connor and Dovers 2004).

Policy cycles and decision process approaches, stemming from work by especially Lasswell (1951), who proposed a 'policy sciences' approach. Decision processes are more narrow and limited to specific issues than the broader and now much more common policy cycle approach, although there are similarities as shown below (Howlett and Ramesh 2003):

Problem-solving phases	*Policy cycle stages*
problem recognition	agenda-setting
proposal of solution	policy formulation
choice of solution	decision-making
putting solution into effect	policy implementation
monitoring results	policy evaluation

Such approaches allow structure in analysing policy and recognise that policy has many parts. The policy cycle model stresses that policy processes are not linear — problems are revisited, policies changed, goals redefined, etc. The problem with 'models' of how policy is 'made' is that most often it isn't made this way at all, but in a much messier fashion. Nevertheless, models of the policy process, if not mistaken for reality or applied too rigidly, can be very useful aids to analysis and prescription.

In most cases, an exercise in policy analysis will evidence a mixture of approaches. This is sensible, as each approach is suited to explaining different aspects of policy — legal issues, political processes, individual choice, group process, etc.

A further consideration is *comparative* policy analysis. With new and complex problems such as sustainability, comparative analysis can aid policy learning by casting widely for lessons. But how to compare across sectors or jurisdictions — on the basis of political or administrative similarity, shared problems, different application of policy instruments, or similar problem attributes? If lessons are sought from overseas policy experiences, do we examine similar political contexts, contrasting ones, or shared issues? If analysis is not comparative, this issue is still important, as it asks where else the lessons of the analysis can be applied. This issue of comparison and lesson-drawing applies also to analyses across policy sectors, for example, public health, community development and environmental management.

Enduring questions in policy analysis

Many questions endure in policy analysis; issues given much attention but not attracting consensus as to resolution or method. It is important that these unresolved areas are understood, and especially that there will be different views on the different approaches such as these:

- the *independence of analysis,* and the relationship between policy analysis undertaken by supposedly independent researchers, and the state or its agencies (or between policy analysis and policy formulation: e.g. Torgerson 1986). Too close an association with government may constrain inquiry, but maximise the uptake of results, and vice versa.
- the tension between *rational-comprehensive* and *incremental* understanding of policy. The former constructs policy as a stepwise, well defined process producing a 'rational' policy outcome. Incrementalism says this is unrealistic — policy-making is complex and politically contingent — and proposes that we proceed in smaller, pragmatic steps (Lindblom 1959). An attractive position is halfway, recognising the messiness and politics of policy but not abandoning the hope of more rigour and direction.
- *the usefulness of models.* The literature abounds with descriptive, analytical and prescriptive 'models' of the policy process, and each will guide the questions asked and methods and data used, being based on particular theoretical, conceptual or philosophical beliefs (see Dryzek, 1997). We should be clear what construction of the policy process analysts subscribe to, as this influences methods and findings. For example, focus on individual choice may ignore institutions and non-economic behaviour, while sociological analysis may overlook economic factors.
- the relationship with *politics, values and the state* (e.g. Brooks and Gagnon 1990). Rational studies that ignore politics are as limited as purely political analysis focusing on immediate conflicts. Some deride 'politics' as expediency that obstructs rational decisions, but policy is inevitably political, dealing with values, rules and allocation of power and resources. The challenge is to impose structure on analysis while factoring in the political context.

- *problem definition*. Without a well defined problem, solutions are hard to find. The policy literature lacks useful typologies of policy problems that extend beyond simple categories (Linder and Peters 1989). Too often, we confuse basic issues (e.g. soil erosion, remnant vegetation protection) with policy problems (e.g. regulatory design, provision of incentives). Rather than classifying issues we can explore generic problem features or attributes (see section 2), and ensure that analysis recognises the loci of policy responsibilities.
- *policy instrument choice*, which is too often a matter of convenience, expediency or disciplinary or ideological bias. Singular instruments are advocated when typically a mixture will be needed, and we rarely observe a full menu of instruments assessed via rigorous selection criteria (see section 3). Across the literature there is little consensus on how to choose instruments or even on what the options are (e.g. Linder and Peters 1989).
- *policy learning* (including policy monitoring and evaluation). Useful learning from policy experiences is 'a concept that is advocated but not adequately conceptualised' in the literature (May 1992: 350), and is not always evident in practice. Learning needs to involve improved understanding and not simply mimicry, according to May, who provides a simple typology:
    - *instrumental policy learning*, about the viability of specific instruments or program design;
    - *social policy learning*, about the social construction of policy problems, the scope of policy or about policy goals; and
    - *political learning*, about political and policy processes and how advocates can better advance their arguments.

Bennett and Howlett (1992) emphasise *who learns*: government instrumentalities; policy actors, or broader policy communities? There is also the issue of whether suitable processes and institutions exist to enable or encourage such learning on the part of agency officials and other stakeholders.

This discussion identifies key questions that recur in policy analysis and formulation, no matter what the issues faced. The next section describes some key features of one set of issues — environment and sustainability — so as to better understand this concern as a set of policy problems, before section 3 provides the basis of a 'language' of policy suitable for communication between researchers, stakeholders and policy agencies in the sustainability area.

2. Sustainability as policy problem

Traditionally, environmental issues were attended separately, in a reactive fashion. Resource scarcity and allocation in particular sectors, end-of-pipe pollution control, and limited nature conservation dominated early thinking and policy, with little coherence as a policy field and without a proactive, long-term perspective. From the 1960s, concern for the 'whole problem' emerged, voiced at the international level in Stockholm through the 1972 United Nations conference on the human environment and stated clearly as 'sustainability' in the 1980 World Conservation Strategy. In what is widely known as the Brundtland report (WCED 1987), the World Commission on Environment and Development stated the modern idea of 'sustainable development'. This led to the 1992 UN Conference on Environment and Development and its *Rio Declaration,* Agenda 21 and related conventions, and more recently to the 2002 World Summit on Sustainable Development in Johannesburg.

Sustainability stresses the interdependence of environment and development, the global importance of environmental and resource issues, and the long-term nature of the issues. Above all, it combines environmental concerns with issues of economic and human development and governance. Most countries have articulated this agenda through domestic policy. The core policy agenda, from the *Rio Declaration* but often restated since, includes the following:

- balancing of inter- and intra-generational equity, providing for human needs now while conserving resources and opportunities for future generations;
- elevating the protection of biodiversity and important ecological processes from marginal to high priority policy goals;
- recognising global dimensions and interdependence in environment and development issues;

- integrating environmental, social and economic issues and policy (the 'integration principle');
- adopting precautionary policy approaches in the absence of scientific certainty when serious or irreversible environmental degradation may occur (the 'precautionary principle');
- recognising the need to involve the broader community in policy debate and formulation and environmental management; and
- recognising the need for new, innovative policy and management approaches, including incentive mechanisms, institutional change, and community-based approaches.

Sustainability presents much more significant challenges to researchers, policy-makers and communities than consideration of single environmental issues. Major sustainability issues include the goals of biodiversity conservation, climate change, human development for the poor, protecting fisheries and oceans, integrated management of land and water; and problems include climate change, excessive consumption by the rich, inefficient use of materials and energy in industrial production, and population–environment issues. Taking biodiversity as an example, we can consider how, from a policy perspective, the challenge has evolved. Past policy responses were limited to creation of a few reserves and protection of notable species. There were few links to other (non-reserve) land uses, and integrated protection of biodiversity across genetic, species and ecosystem levels was hardly attended. We now realise that protection of all species, in and out of reserves, is necessary, that to do so requires protection of ecological processes such as hydrological and nutrient cycles, and that management of threats to biodiversity (such as farming practices, climate change, feral species) involves links to other areas of policy. This reframed problem set requires different forms of information, different policy approaches, and different structures and processes in the institutional system such as connection between conservation, agricultural and economic agencies.

To help approach sustainability as a set of policy problems to solve rather than simply issues to debate, we can identify the *attributes* of these problems that present policy challenges. Major issues in sustainability, when considered in this way, display some attributes more often, and more often in combination, than many other policy problems (Dovers 1997):

- broadened, deepened and highly variable spatial and temporal scales;
- absolute ecological limits to human activity;
- irreversible impacts, and related policy urgency;
- complexity within and connectivity between problems;
- poor information, and pervasive risk and uncertainty;
- cumulative rather than discrete impacts;
- important assets not traded in formal markets and thus rarely given economic value;
- new moral dimensions (other species, future generations);
- 'systemic' problem causes, embedded deeply in patterns of production, consumption, settlement and governance;
- difficulty in separating public and private costs and benefits;
- lack of uncontested research methods, policy instruments and management approaches;
- lack of defined policy, management and property rights, roles and responsibilities; , and
- strong demands and justification for increased community participation in both policy formulation and actual management.

This is a long list and paints a complex but realistic situation for policy-makers. Not every attribute will be relevant every time. But even when only two or three of these attributes are strongly evident in a specific situation, sustainability problems may be *different in kind* from those in other policy fields (e.g. service delivery, social policy). This suggests we might need to change policy processes, institutional arrangements and modes of analysis. Other policy problems are not 'easy', but existing analytical and policy approaches, and the institutional settings these emerge from, have been developed by reference to longer-standing concerns that do not so often display these attributes. New and different problems arguably will require new and different mechanisms.

These attributes can be used to define what the main policy challenges are for a specific issue, as the importance of particular attributes will vary from case to case. They can also be used to allow greater specificity about the contribution that can be made by particular disciplines (e.g. those described in this book) and other knowledge systems (e.g. local, Indigenous).

Drawing on sustainability principles and existing policies, and on these attributes, we can identify the key demands sustainability places on policy:

- improve information capacities (gathering, manipulation, communication);
- improve policy and management coordination and integration across policy sectors and portfolios, and across jurisdictions;
- increase longevity and persistence in policy processes and initiatives;
- enhance policy learning across space and time;
- improve capacities and techniques for policy analysis and policy choice;
- enhance community participation in policy and management; and
- improve institutional and organisational capacities to achieve the above.

These imperatives describe a difficult task, and serve both to explain the widely recognised failure of policy and institutional responses to the sustainability challenge so far, and to suggest ways to formulate more effective responses. Section 3 presents a framework for policy analysis and prescription that can assist in addressing these tasks.

3. A policy checklist for sustainability

Figure 6.1 presents a detailed view of elements of effective policy processes for sustainability. This is not a 'model' to be strictly followed, or a representation of what actually happens, but a checklist and framework to inform analysis and prescription – reminding us of things that should not be forgotten. Some elements may seem obvious, but experience shows they do get left out. The elements need not happen in a linear fashion, but are interconnected and cyclic. Not all elements must be met in each case; a particular policy may be informed by, say, discussion of social goals or environmental monitoring serving a range of policy programs, or made viable through existing administrative and legal settings. But it forces the question of whether these elements have been attended to, and whose responsibility they are. Some elements will be more or less important in different situations. The emphasis is too often largely on the policy statement, or 'avowal of intent' (element 10 of the model), with perhaps some reference to implementation, at the expense of what must come before and after.

Many elements are more often than not inadequate or missing, in much sustainability policy. Inadequate basic environmental monitoring, weak legal and institutional provisions, under-resourcing, and poor policy evaluation are obvious cases. A crucial point is that inadequacy in *one element* may cause policy failure – for example, lack of a defensible legal mandate, insufficient resources, or poor communication. This makes the checklist a valuable tool. Attention to different elements or aspects of the full 'process' of sustainability policy demands input from different disciplines, professions, parts of community, skills and methods – policy analysis as interdisciplinary field. For example, the following disciplines and professions can inform thinking about selected elements of the framework in figure 6.1 (as well, of course, as members of the broader public):

Monitoring environmental change, problem identification (elements 3–5, figure 6.1) – information sciences, demography, ecology, hydrology, law, economics, geography, psychology

Identifying uncertainties (element 6, figure 6.1) – philosophy, information sciences, mathematics, statistics, ecology and other natural sciences, public policy, law, political science

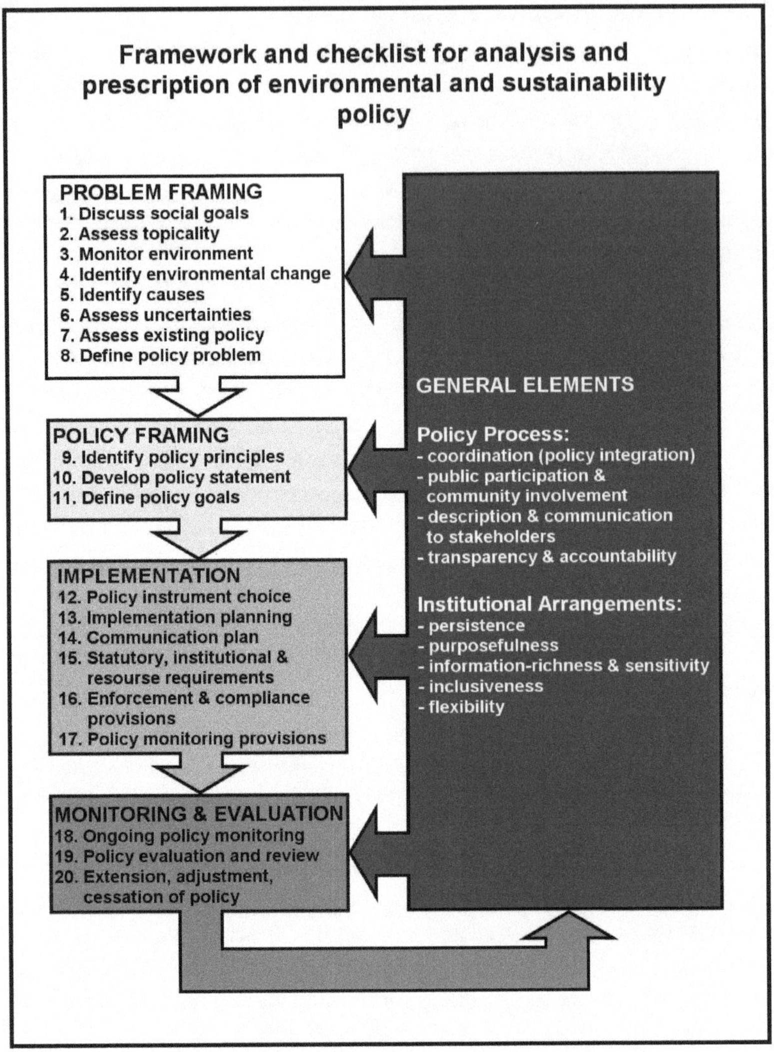

Figure 6.1. Checklist for environmental and sustainability policy.

Assessing existing policy (element 7, figure 6.1) – economics, public administration, public policy, law, history, planning, institutional theory

Implementation planning and requirements (elements 13–15, figure 6.1) – public administration, law, public policy, accounting, geography, sociology, psychology.

Drawing on a range of the different approaches to policy analysis identified earlier is also necessary for explaining and undertaking the different elements of sustainability policy.

A glimpse into greater detail

Although figure 6.1 appears complicated, it only hints at the complex reality of policy processes – each element is the subject of an extensive body of literature and requires considerable expertise to properly understand, let alone undertake effectively. To illustrate such detail, two elements can be explored further: element 12, policy instrument choice, and the general elements of public participation and institutional arrangements. Important in themselves, these also provide contrasting views of policy – instrument choice is somewhat 'top-down' from a government perspective, and public participation is 'bottom-up' from the perspective of the community– and the broader institutional settings that allow both approaches. Similar and much greater detail could be given for all other elements in figure 6.1.

Policy instrument choice is a crucial matter, but too often choices are made on the basis of convenience, disciplinary bias or familiarity rather than informed judgement. This is unsatisfactory – no one instrument will always be the best, many other options exist, and within such general classes there are many specific instruments. The main classes of policy instruments are:

- research and development, monitoring
- communication and information flow
- education and training
- consultation and mediation
- intergovernmental agreements and conventions

- statute law (legislation)
- common law
- covenants on property titles
- assessment procedures (e.g. environmental impact assessment)
- self-regulation
- community involvement
- market mechanisms (economic instruments)
- institutional and organisational reform
- change in other conflicting policy settings
- inaction (with reason).

This greater detail does not make choice easier — in fact it makes it harder — but it does force a more reasoned and informed basis for choosing policy options. However, each class of instrument contains a variety of more specific options, as the following examples show:

Education and training: public education (moral suasion), targeted education, formal education (schools, higher education), training (skills development), education regarding other instruments.

Statute law (legislation): new statutes or regulations under existing law to create institutions, establish statutory objects and agency responsibilities, set aside land for particular uses, do land use planning, control development; enforce standards; prohibit practices.

Market mechanisms: input/output taxes/charges; use charges; subsidies; rebates; penalties; tradeable emission permits/use quotas; tradeable property/resource rights; performance bonds; deposit-refunds.

Faced with so many options, criteria for selecting policy instruments need to be applied so that the option chosen suits the problem at hand. Selection criteria for instrument choice can be divided into those that will influence effectiveness, and later implementation:

Effectiveness criteria: information requirements; dependability (to achieve goals); corrective vs. antidotal focus (address symptoms or causes); flexibility (across contexts, time); cost and efficiency; influence on other policy sectors; equity impacts.

Implementation criteria: political/social feasibility; legal/constitutional feasibility; institutional feasibility; monitoring requirements; enforceability/avoidability; communicability to those affected.

So policy instrument choice is far from a simple matter. This should be no surprise, as policy interventions to enhance sustainability are intended to change the behaviour of individuals and groups operating in complex social systems, interacting with complex natural systems. Such interventions need to take this complexity seriously and choose the means accordingly.

Importantly, rarely is a single instrument used in implementing a policy — a mix of instruments is required even when one instrument represents the chief strategy. For example, a price mechanism such as a tax will require a legislative base and some form of communication or education. If we think of all instruments as forms of *information*, intended to change the behaviour of individuals and groups, then policy instrument choice can be approached as the task of designing the mix of media for the message. Three classes of instrument, though, will always be required, even if as minor elements. These are *legal*, in societies subject to the rule of law; *institutional*, because humans only achieve collective goals through institutions; and *educational*, when people need to know what is going on. This last point brings us to the second of our three glimpses into greater complexity.

Public participation and community involvement are options that may be chosen by policy networks to achieve goals, as well as being options demanded by many people. Public participation is a principle stated in sustainability policy, being viewed as ethically desirable as well as useful in ensuring effective policy. Often there is poor understanding of the many forms of participation beyond general notions of democratic inclusion in the political process and community participation in on-ground environmental management. These are some of the main participators and forms of public participation in politics, policy and management for sustainability:

- voters in elections
- individual lobbyists or members of interest groups such as

conservation organisations, farmer groups, political parties, consumer associations

- consumers, through informed choice (e.g. eco-labelling)
- workers in industries, trades and professions implementing environmental practices
- recipients of information about policy choices, changes or implementation, or providers of information, or participants in research projects that will inform policy
- statutory rights to know, rights to object to developments, legal standing in courts or through environmental or social impact assessment
- mediation or conflict resolution processes
- through commissions of inquiry, parliamentary committees, etc.
- inclusion (directly or through representatives) in short term policy processes
- advisory or management boards or committees
- community-based monitoring or resource and environmental management groups
- community based or cooperative management arrangements (co-management), where actual management responsibilities are devolved.

So there are many options. There are also different *purposes*: to gather or disseminate information; advocate policy change; enforce transparency and accountability; implement policy; or undertake management works. What comprises 'the public' or 'community' is often not clear, as human societies are made up of countless communities defined by location, demographic factors, cultural or religious similarities, employment, recreational interests, and so on. Considering the purposes and kinds of participation, along with the subsets of the community with an interest in the policy in question, allows greater sophistication in designing and implementing policy, and in informing citizens' access to the policy system.

Both instrument selection and public participation in policy will be determined by *institutional arrangements*, the third of our glimpses into the greater detail behind figure 6.1. Most discussions of sustainability state that our current institutions must change — this is not surprising, as present institutions are products of past understanding that did not include much thinking about sustainability. In figure 6.1, five principles are stated that describe features of institutions more likely to advance sustainability, and these are described a little further below:

Persistence, allowing sufficient time for policy and institutional 'experiments' to be run and lessons learned. This is particularly important given the time it takes for policies to be fully implemented and the result to become apparent, when compared to short term political agendas;

Purposefulness, through being based on a set of core policy principles that are clearly stated and widely understood (such principles will vary according to the context, but high level principles are stated in international sustainability documents such as the United Nations' 1992 *Rio Declaration*);

Information-richness and information-sensitivity, especially over time, referring not only to information-gathering but to its wide ownership and application among multiple stakeholders;

Inclusiveness, through clearly understood, sustained public participation in both higher level policy and on-ground management; and

Flexibility to encourage adaptation and improvement, and to prevent persistence and purposefulness from developing into rigid and unchanging positions.

These principles draw on both theoretical discussions and on observations of what has been more or less successful with actual organisations and institutions (discussed and applied in Dovers and Wild River 2003). As with all the topics touched upon in this chapter, these five principles are only one, summary description of what suitable institutions for sustainability would look like.

4. Conclusion

For many people, 'policy' is a dull and boring topic. In many ways, that is true: the detail of policy processes are complex, often opaque, and certainly not as photogenic as other topics in environment and sustainability. Many other aspects of environment and sustainability are more engaging, as they deal with immediately pressing issues (e.g. poverty, climate change) or inherently interesting and appealing things (e.g. rivers, wildlife). Nevertheless, these more engaging things cannot be fully comprehended or responded to without some grasp of policy, for it is through policy processes that pressing issues can be resolved. It is difficult to imagine a career or a cause in the environmental field that does not in some way require a person to come into contact with policy, whether as a participant or observer. This chapter has sought to provide some ideas and frameworks to understand the policy domain, and to invite interest in the much greater detail that is available.

5. Further reading

Public policy: a critical approach (Considine, 1994) provides a good introduction to the field of public policy. For terms and definitions, see also *Studying public policy: policy cycles and policy subsystems* (Howlett and Ramesh, 2003).

The original definition and context for sustainability is found in *Our common future* (WCED, 1987). For a conceptual framework and analyses of case studies of sustainability policy, see *Institutional change for sustainable development* (Connor and Dovers, 2004). John Dryzek's *The politics of the earth* (1997) suggests ways to explore further the different 'discourses' around an issue such as sustainability, identifying the underlying assumptions that give meaning to different interpretations.

For an expanded discussion of the elements in figure 6.1 and other issues raised in this chapter see *Environment and sustainability* (Dovers, in press).

7 ECOLOGY

D.L. Hazell, R.G. Heinsohn and D.B. Lindenmayer

1. What is ecology?

Ecology is a discipline that attempts to understand and explain interactions between living things and their environment. It draws from many disciplines in examining these interactions, integrating botany, zoology, earth sciences, physical sciences and geography, depending upon the subdiscipline of ecology or particular interactions of interest. While ecology is often nested within universities under the umbrella of botany or zoology, the relationship between these disciplines is far from hierarchical (see figure 7.1). To describe oneself as an ecologist leaves much to the imagination — subdisciplines such as plant, behavioural, landscape, community, population and evolutionary ecology can differ greatly in their respective methods, concepts and underlying research philosophies.

2. Where has ecology come from?

Homo sapiens and our ancestors have always interacted with and relied upon the environment for survival. Many Indigenous peoples demonstrate a well developed understanding of ecological processes and relationships that has developed over thousands of years, through cultures of respect for the natural environment. It is important to emphasise, however, that any new settlers in a new land will always have a period of adjustment as they learn what works and what does not. Some cultures have been late in recognising the pivotal importance and fragile nature of the natural environment. Self-conscious Western ecological thinking has only emerged during the last few centuries.

Entire books have been written on the history of ecology and ecological thinking. From these it is clear that there is no distinct single lineage of development for ecology as a modern discipline. In addition, historians rightly argue that early ecological ideas do not reflect the development of a discipline. Only with the value of hindsight can we see which ideas have carried through to modern ecology. A few notable historical events must serve here to demonstrate the breadth of approaches and philosophies of environmental interest which inform ecology.

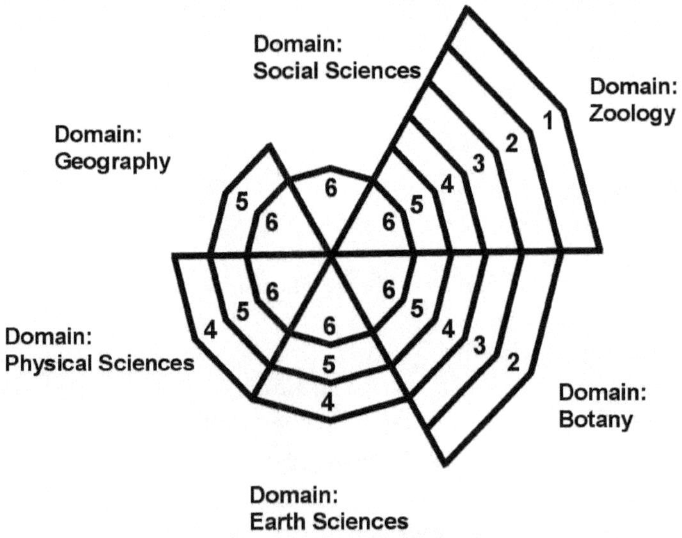

KEY TO RELATED AND SUB-DISCIPLINES OF ECOLOGY

1. **Animal behavioural ecology**
2. **Plant or animal population ecology**
3. **Community ecology**
4. **Freshwater ecology**
5. **Landscape ecology**
6. **Human ecology**

Figure 7.1. The various domains of related disciplines and subdisciplines of ecology.

Natural history was one of the precursors of modern ecology, promoting an understanding of the environment through personal experience. Gilbert White (1720–93) is considered one of the founders of modern respect for nature. His *The natural history of Selborne*, published in 1789, is one of the most republished books written in the English language (Griffiths 1996). White studied the plants, animals and surroundings of his village of Selborne and wrote with amazement of the relationships he observed in nature. The context for the work of this early naturalist was the highly modified hollow lanes of Hampshire and it was here he observed 'nature' and its interrelationships.

The late 17th century and the 18th century saw a growing interest in natural history. Wealthy Westerners collected natural history specimens, and there was great interest in describing and classifying plants and animals on the basis of relationships and physical similarities. Outstanding in this pursuit was the Swedish botanist Carl von Linné or Linnaeus (1707–78) whose *The oeconomy of nature* represents an early attempt to organise and characterise ecological interactions. These interactions were presented as static phenomena with a defined space and clear role for every creature and a strong anthropocentric and creationist (i.e. the hand of God) perspective. The term 'oeconomy' was used as far back as 1530 in reference to household management and was also applied later to the political administration of a state's or community's resources.

Many early efforts to understand the natural environment were caught in the crossfire between creationists and evolutionists. The work of Charles Darwin (1809–82), especially *On the origin of species* (1859), was central to evolutionary theory. His notion of natural selection was, however, rejected for the most part by many other scientists at the time because of the lack of a defined mechanism (later clarified by Mendel's work on genetics). While Darwin's arguments focused on evolution, many of his ideas shed light on ecological relationships. His work developed early ecological ideas such as population stress, resource limitations and the notion that the physical environment species inhabit is not static.

In the writings of field ecologist and transcendental philosopher of nature Henry David Thoreau (1817–62), we see the emerging contrast between reductionist and holistic approaches to studying natural relationships. Like Gilbert White, Thoreau shared a strong (Arcadian) respect for nature and felt humans had no right to exploit or manipulate nature for their own purposes. One of the first to contemplate ecological succession, Thoreau was troubled by the detached, cold way in which science engaged with nature, and stated: 'The inhumanity of science concerns me, as when I am tempted to kill a rare snake that I may ascertain its species. I feel that this is not the means of acquiring true knowledge' (quoted by Worster 1994: 91). He strongly disliked the failure of science to accept other forms of inquiry which were considered worthless or untrustworthy. Thoreau would have been sympathetic with modern branches of ecology that embrace the integration of diverse forms of inquiry, such as landscape and human ecology.

Scientists, philosophers and natural historians were attempting to explain and understand ecological phenomena well before an actual discipline emerged. The emergence of ecology did not reflect changing ideas about nature. Rather, it occurred once a community of scientists developed who saw the study of natural relationships as their main objective. It became evident that field studies were required to answer certain biological questions. However, there was a strong desire to transform field studies into genuinely scientific endeavours that shared the rigorous approaches to measurement and quantification of the 'hard sciences' such as physics. This resulted in ecology – described as a fusion of physiology and natural history (McIntosh 1985).

The term 'oecologie' was coined by Ernst Haeckel in 1866, with obvious links to the already established concept of oeconomy. It was not until the 1890s that ecology gained recognition as a major division of biology. Ecology was far from a single discipline. From backgrounds as diverse as plant physiology and oceanography, scientists examined relationships between organisms and their environment under the umbrella of ecology, using a variety of theoretical models. A diversity of philosophies and agendas continued to drive the development of ecology into the 20th century. This included utilitarian versus Arcadian ('back to nature') influences as well as holistic field studies (with a strong natural history influence) versus reductionist, laboratory-based, 'hard science' approaches. With such a complex of ideas and philosophies it is no wonder debate continues over the true history of ecology, its founders and even what ecology is. It is also not surprising that modern ecology is plagued by uncertainty about the definition of terms that are central to the discipline and the scope of questions that ecologists attempt to answer. Uncertainty is a major difficulty for ecology as it demands ongoing questioning of the very fundamentals of the discipline.

The relationship between ecology (the science) and conservation of the environment warrants mention in a historical context as it contributed significantly to the prominence of popular versions of ecology from the 1960s onward. This relationship also underpins some of the philosophical approaches to ecology that still exist. The utilitarian philosophy that underpinned early developments in ecology left the natural environment open to exploitation. Many early ecologists, such as American grasslands ecologist Frederic E. Clements (1874–1926), were motivated by a need to understand natural systems so they could be more effectively used by humans. During these early days of development of the discipline there were no strong links with the environmental movement (Bowler 1992: 309). However, by the 1950s concern for the destruction of the natural environment sought out ecology as a means of providing scientific support for conservation. Practical forester Aldo Leopold (1886–1948) argued against human use of nature as a mere commodity and promoted ideas of a land ethic. Entire societies started to express these views in the 1960s with those such as Rachel Carson (1907–1964) highlighting the effects of pesticides on the environment in *Silent spring* (1962). In 1966 zoologist Alan J. Marshall raised concerns about reductions in biological diversity in *The great extermination*. Ecology was thrust into the limelight as the defender of nature and a discipline of science that could educate the masses about land degradation and the importance of the environment. Ecology became political as well as scientific. Scientists who wished to engage in the conservation debate found themselves explicating not just science, but also social responsibility and moral imperatives. While ecology has been influenced by the hard sciences in developing certain methods and philosophies, it is clear that it does not sit easily within the classic hard science model. However, plant ecologist David Ashton distinguished between ecology, conservation and environmentalism in 1993 when he declared that 'ecology is the study of why plants and animals are where they are ... Conservation is an appreciation of what we have and want' (quoted in Robin 1998: 71).

Many streams of ecology now focus on contributing to conserving nature for its own sake, representing a considerable shift from some early ecology which saw the science in the service of the human exploitation of nature (Bowler 1992).

3. Key concepts and world views

Fundamentally ecology is a science and follows a number of scientific methods as its mode of inquiry. One of the most well known scientific methods of reasoning is the hypothetico-deductive approach espoused by Karl Popper. Presenting a hypothesis and attempting to falsify it in the Popperian way is not always possible in ecology, where in many cases the laboratory is the real world and variables cannot be held constant. The complexity of ecological systems requires the use of multiple methods of reasoning which can include inductive reasoning (evidence drawn from particular cases to generate general statements), construction of causal explanations, use of contrasting questions and consideration of multiple postulates (claims which form the basis of reasoning).

Ecology is assisted by disciplines such as taxonomy, chemistry and genetics that name, group, characterise, split and describe basic building blocks upon which our understanding of the natural world is premised. Ecological concepts may be grouped into three different categories: natural, functional and integrative. Natural concepts relate to organisms or features of the environment (e.g. a bird, a tree, fire or rain). Functional concepts consider measurable process, interaction or structure that apply to natural concepts (e.g. predation, migration, competition, photosynthesis). Integrative concepts consider the organisation or properties of an ecological system (e.g. a population, community or an ecosystem; or resilience or diversity). Integrative concepts cannot be measured directly and require theory and the use of functional concepts for explanation. We present some of the most common ecological concepts below but emphasise that they are merely concepts (ideas that are defined in order to facilitate understanding). They provide explanations of the natural environment that are widely recognised.

While interactions may occur at a cellular level the notion of an individual *organism* is explicit in the definition of ecology as the fundamental building block. Individuals from the same *species* collectively may constitute a *population*. Interactions between individuals from the same population may include *breeding* and *competing* for *mating partners, habitat* and *food*. While functional concepts such as reproduction and competition connect individuals from the same population there is no universal definition for a population that results in the same use of the term in every circumstance. Begon et al. (1996) define population as 'a group of individuals of one species in an area; though the size and nature of the area is defined, often arbitrarily, for the purposes of study being undertaken'.

When populations of different species interact by co-occurring in space and time they are known as a *community*. A community consists of the sum of its species components plus the interactions between them (emergent properties). These may include *competition* for food and habitat, *predation, parasitism* and *mutualism* (where two species benefit each other through their relationship). Communities may be defined at any spatial scale within a hierarchy of environments. They may be based on a particular taxonomic group or activity (such as herbivory). Certain communities may be nested within others that are more broadly defined (e.g. an aquatic invertebrate community may be nested within a temporary wetland community). On newly exposed landforms (e.g. after a volcanic eruption) or after disturbance, *plant* communities may undergo *succession*. In such cases plant community composition changes over time. Early colonisers gradually change the environment and make it more suitable for other species until the community eventually reaches a stable endpoint or climax state that is largely determined by climate. There are other alternatives to the notion of a climax state, such as vegetation communities reaching equilibrium with surrounding influencing factors and the notion that biological legacies (such as trees, logs, seeds and eggs) left behind by a previous disturbance event (like a fire) shape the trajectory of the recovery process (Franklin et al. 2000). These sorts of ideas contributed to the development of one of the most holistic concepts in ecology – the *ecosystem*. This consists of a complex of organisms (community or *biome*) and their inorganic environment existing in a dynamic equilibrium (Tansley 1935).

Space and *time* are both central considerations in the practice of ecology. They shape the dynamics of relationships between organisms and their environment. For example, the availability of habitat may vary over space, influencing the *distribution* and *abundance* of a particular species. Such dynamics may reflect variability in characteristics such as soil type, climate, elevation or more anthropogenic influences (e.g. land clearing). The availability of inhabitable environments may also vary over time. For example, climate change could result in the loss of a particular forest type as conditions become unsuitable for tree survival or reproduction. Habitat availability may also increase or fluctuate over time, depending upon the nature of the underlying processes, which could include drought, fire or seasonal variation in food availability.

4. Theories of ecology

The development of ecological theory has centred around six themes (as identified by Walker 2002): the distribution and abundance of organisms; predator–prey dynamics; biological diversity and species richness; community and ecosystem dynamics; applied ecology; and complex adaptive systems. Here we discuss just a few key concepts.

Ecologists may seek to explain the diversity of species in various communities by either equilibrium or non-equilibrium theories. An equilibrium theory focuses on the properties of a system at an equilibrium point, or point of balance, whereas non-equilibrium theories are concerned with perturbations and variation away from equilibrium. A famous example of the former is MacArthur and Wilson's (1967) equilibrium theory of island biogeography that states that the number of species on an island is determined by a balance between *immigration* and *extinction*. Species continually go extinct and are replaced by the same or different species. The exact rate of immigration depends on the remoteness of the island from its source of potential colonisers, and the rate of extinction depends on the size of the island (big islands hold more species), the number of species already there, and the extent of competitive exclusion. MacArthur and Wilson's theory has been used to explain species richness on real oceanic islands as well as islands of habitat created by human disturbance.

Ecology is becoming more and more concerned with the effect of disturbance to ecosystems but distinguishes between regular disturbances such as tidal movements, and other less frequent events. *Disasters* happen frequently enough to exert selection pressure on organisms over time. Some plants, for example, have adapted to serious fires even if they are relatively infrequent. *Catastrophes* on the other hand are disturbances, such as volcanic eruptions, that happen too infrequently for adaptation to occur. *Resilience* is a measure of the ability of the system to absorb changes and still persist. This includes the speed at which it returns to its former state. *Resistance,* on the other hand, describes the ability of the community to avoid disturbance in the first place. *Dynamically fragile* systems are those that are stable within a narrow range of conditions whereas *dynamically robust* systems are stable over a wide range of conditions. The conventional wisdom has been that complex communities are more stable and less prone to disturbance, but this is not borne out by recent theoretical and empirical analyses. For example, fragile complex communities in stable environments are among those that could be most prone to disturbance by humans.

5. Methods

Ecology involves the collection and analysis of observations in order to generate an understanding of the natural environment. The individual ecologist must make many (subjective) choices in this process, revealing the art as well as the science of ecological research. The nature of these observations, where they are collected from, and how they are collected and analysed can vary enormously. There are major issues in striving for consistency and comparability across ecological studies. A range of reference texts has been developed in an effort to assist with this challenge.

Ecological observation may consist of measured numerical data (quantitative data) or descriptions of the qualities of the subject of interest (qualitative data). These characteristics of observations influence how they may be summarised and analysed. For example, highly qualitative data may be analysed using inferential statistics. This allows generalised conclusions to be drawn beyond the actual observations made.

Ecological studies can be undertaken in the laboratory, or through field-based research. This is a major distinction, often reflecting differences in philosophy, methods of reasoning and subdisciplines of ecology. Laboratory-based research allows a greater focus on experimental work. This involves treatments that are deliberately imposed by the researcher in order to establish effects. Causality is readily established through experiments, as the researcher can manipulate one element in a system while ensuring that all other influences within the laboratory environment remain constant. Field-based ecological research can be experimental but is more often observational as it is difficult to manipulate certain parts of the natural environment while also maintaining a constant (control) environment for comparison. Causality is difficult to establish in observational studies as many variables (measured and unmeasured) could be the ultimate cause of the observed differences.

Laboratory-based ecological studies encounter the difficulty of replicating the complexity of natural systems and may simplify natural phenomena and relationships and study their component parts. Such a reductionist approach is limited in its capacity to identify emergent properties of ecological systems that result from their complexity, such as those associated with communities (see section 3). Reductionist laboratory-based experiments can, however, provide a useful framework for following up observational, field-based ecological studies that identify ecological relationships, but are unable to demonstrate causality. The need to incorporate complexity into ecological experimentation is reflected through the use of microcosms, where natural systems are set up under laboratory conditions. Large 'natural experiments' also represent recent attempts to overcome the limited ability to draw strong inference or causality from field-based observational studies. Such experiments involve manipulation of geographic areas through treatments such as fire or tree clearing.

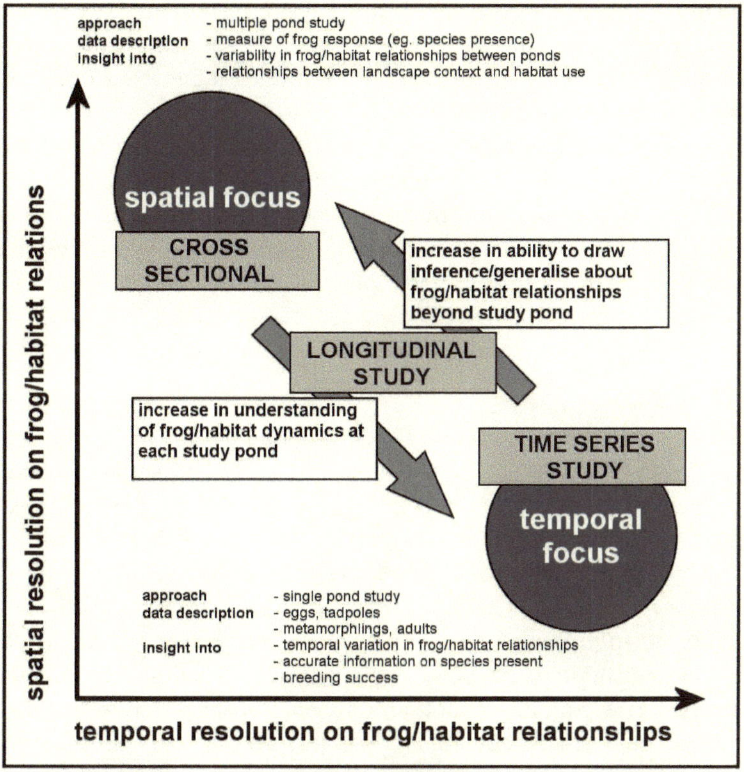

Figure 7.2. A comparison of data strengths and weaknesses between a single pond study with a temporal focus on frog/habitat relationships and a multiple pond study with a spatial focus on frog/habitat relationships. The ability to draw inference beyond ponds that are examined increases as more ponds are included in the study. In contrast, the ability to understand detailed dynamics of frog/habitat relationships at each pond decreases as more ponds are included in the study.

Field-based studies allow insight into spatial and temporal influences on ecological phenomena. Trade-offs are required, however, as both cannot necessarily be achieved through the same design. Choices of experimental design, survey techniques and survey effort are decisions that determine the strengths and weaknesses of the resultant data. Figure 7.2 provides an example of the challenges of designing an ecological study in order to understand frog/habitat relationships. Several different study designs are represented. Cross sectional studies have many independent single outcomes (i.e. many observations across space). Such a design allows inference or generality regarding habitat relationships across space. In a time series study there are many repeat measures made over time, often at one locality. This provides insight into habitat relationships over time but limits the capacity to draw inference beyond the study observations. Longitudinal studies represent the middle ground between these two study designs, consisting of multiple single outcomes measured repeatedly through time (figure 7.2).

6. What ecology is not

Ecology has a clear definition, but through its involvement in recent conservation debates it is now perceived as an endeavour encompassing all things environmental. As a term, ecology has been borrowed by other disciplines and applied in many different contexts that serve to confuse. For example, human ecology is the study of interactions between organisms and their environment including humans as an organism, while considering economic, social and political organisation as part of the environment. The link between the discipline of ecological economics and relationships between organisms and their environment is not so clear. In this context ecology is applied in order to imply a more holistic approach to economics (see chapter 4). Usage of the term ecology occurs across many disciplines.

The distinction between ecology and conservation is important to emphasise. Ecology is often practised with no connection to conservation, and vice versa. While ecology can contribute knowledge on which to base conservation decisions, it does not make conservation happen. Incentives, communication, education and/or cultural change are more relevant in this situation. An understanding of these phenomena sits well outside the discipline of ecology.

Conservation biology is a relatively recent and emerging discipline. Although it encompasses many things, its primary and overarching goal is the application of ecological principles in the conservation and maintenance of biodiversity. It aims to do this both by identifying areas such as reserves set aside primarily for conservation, and by encouraging ecologically sustainable management of natural resources (e.g. forests, fisheries, agricultural land). Conservation biology can be a multidisciplinary area of study that — while based broadly on ecological thinking — also attempts to link with other fields such as statistical science, climatology, mathematics, social science, economics, and politics. It also involves drawing on fields within or closely allied with ecology such as behavioural ecology, evolutionary ecology and genetics.

7. Principal contributions to furthering our understanding of the environment

Ecology has improved our understanding and appreciation of how the natural world functions. It has highlighted the importance of the natural environment in supporting humans as a species, the amazing intricacies of life but also the magnitude of destruction humans have caused. Ecology has promoted the notion of interdependency, flow-on and indirect effects in natural systems. Simple but highly valuable ideas such as food chains have emerged from ecological thinking.

While ecology provides a framework for examining intact natural systems there are many related or subset disciplines that focus on understanding, minimising and ameliorating the impacts of human change. Disciplines such as restoration ecology and conservation biology have developed rapidly through an urgent need to save poorly understood organisms and systems from extinction or further degradation. Ecological knowledge is central to improving our capacity to manage and use natural and modified systems in a sustainable way. The influence of ecology is evident in concepts such as agroecosystems — the application of ecological systems thinking in agricultural environments.

Ecology also underpins many management approaches designed to solve natural resource management problems. For example, ecological concepts such as resource and habitat availability, community dynamics and ecosystem dynamics have been applied to problems such as nutrient fluxes in waterways and associated blue-green algal blooms. Management strategies for reducing the occurrence of blooms have been developed through an understanding of the relationships between blue-green algae and its environment. Ecological concepts also underpin many management strategies designed to reduce the impacts of feral animals and weed species, in both natural and agricultural systems.

We are still far from understanding the natural world in its intact state. Perhaps, the greatest contribution that ecologists can make is to emphasise how much we don't know and how little capacity we have for estimating the impacts of further destruction of the natural environment.

8. Controversies

There are many controversies within ecology. Concepts such as populations, communities and ecosystems are widely recognised, even beyond ecology. However, the characteristics of these phenomena, influencing factors, dynamics and internal workings remain hotly contested in a theoretical context. Some major examples include the role of equilibrium in natural systems, the value in examining ecological systems from a holistic versus reductionist perspective, and the need for ecology to develop theoretical generality. The value of theoretical versus empirical ecological work has also been debated. However, further integration of the two will improve the effectiveness of ecology as a basic and applied science (Pickett et al. 1994).

Controversy is also widespread in applied ecological research. Biological diversity or 'biodiversity' emerged in the late 1980s as a central paradigm in nature conservation. It encompasses all of nature's variety including ecosystems, ecological processes, species of plants, animals and micro-organisms and their genes. Biodiversity has developed currency within political, community and scientific spheres and has facilitated communication across these spheres about nature conservation. One of the most appealing, but also controversial, aspects of biodiversity as a conservation tool is that it can be measured. Biodiversity may be measured through a simple count of the number of species in a given area. However, there are many other approaches including consideration of gene richness, assemblage richness, land class richness or willingness to pay for future potential value. With such a diversity of approaches, the biodiversity of a particular area may be rated differently, depending on how it is measured. The role of biodiversity in driving nature conservation raises challenging ethical questions. Should every species be considered equal? Are our reasons for conserving biodiversity primarily anthropocentric in nature? If so, how does this influence the way in which biodiversity is assessed?

Biodiversity has proved to be a good common use descriptor for the diversity of nature. However, its value as a conservation tool requires ongoing assessment. There are several related topics that are just as strongly debated. These include the value of indicator species — taxa considered to be surrogates for the presence or abundance of other species (that are difficult to measure) —and the use of rapid biodiversity assessment to provide a quick but reasonable estimate of all elements of biodiversity within a given area.

'Ecosystem services' is a recent concept developed to assist with natural resource management decisions. It places dollar values on ecological processes. When ecosystems are altered there is often a loss in the 'services' the ecosystem provides, such as water or air quality. The costs and benefits of ecosystem services are quantifications of the ecological processes directly supporting human welfare. Conversion of ecological losses into dollar terms may be a powerful tool for the protection of natural ecosystems. However, a focus on ecosystem services could lead to trade-offs in terms of what species we are prepared to lose in relation to their role in ecosystem function. Because it uses dollar value, the concept of ecosystem services favours costing ecological processes that are relevant to supporting human welfare, as opposed to the welfare of other species. There is a likelihood that decisions based on ecosystem services will favour anthropocentric conservation.

Ecology has much to contribute to environmental decision-making. One of the greatest challenges is to overcome the apparent mismatch between ecological processes and values and the short political time frames and monetary measurements of value that influence such decisions. While there is a paucity of ecological information upon which to base management decisions, this does not stop decisions from being made. Ecologists who wish to engage in natural resource management must balance the need for stronger theory with the urgent need for ecological knowledge.

9. Further reading

For a general summary of ecology that fosters common disciplinary goals, see Allen and Hoekstra (1992); for a comprehensive conservation biology text, see Burgman and Lindenmayer (1998); for background on the concept of ecosystem services, see Costanza et al. (1997); for an outline of various ways of measuring biodiversity see Humphries et al.; and for an outline of the inadequacies of ecology see Peters (1991), who argues that much of ecology is not science.

8 ENVIRONMENTAL EARTH SCIENCE

R.J. Wasson

1. Introduction: origins of environmental earth science

Environmental earth science is the study of the rocks, soil and landforms of the planet, and their interactions with people and society (Merritts et al. 1998). The explicit inclusion of people distinguishes environmental earth science from earth science. So earth science becomes environmental earth science when people are included as forces of change to soils and landforms, and as the recipients of the effects of those changes. Because environmental earth science has evolved (starting in the 1970s) from earth science, as environmental concerns have grown and earth scientists have struggled to understand their place in this new world, this chapter begins with an account of the intellectual origins of earth science. It then explores the problems now identified as amenable to the approaches of environmental earth science. This is followed by a brief, but hopefully useful, account of approaches and key concepts in environmental earth science. This account reflects both what is and what ought to be, so it attempts to orient the field, not just describe it. A case study of water pollution in Bangladesh is followed by a discussion of the transport of sediments in a river catchment. This section gives insights into both approach and method, but cannot be comprehensive in so few words.

Environmental earth science draws upon geology, soil science, hydrology, and geology. Geomorphology, geology and geography have intellectual origins that are inextricably linked. Geography can trace its origins to the ancient Greeks, particularly Strabo in 18 AD who described the role of the geographer as explaining 'our inhabited world in its size, shape and character, and its relations to the earth as a whole' (quoted in Jones 1917). There are also traces of geographic thought in Kautilya's Arthasastra, believed to have been written at the time of the Mauryan Empire of northern India in about 300 BC (Kangle 1965). Strabo's account of geography includes both human and non-human elements of our planet, a concern which has enlivened geographers ever since.

The beginnings of modern thought in both geomorphology and geology can be traced to James Hutton (1795). Hutton saw that the Earth was old and that landforms took a long time to develop by processes that are slow. He recognised that landforms are the result of uplift of rocks, erosion (also called denudation) of the uplifted rocks, deposition of the eroded products to form new rocks by lithification (the production of hard rocks from soft sediments) which are then uplifted, and so on. Hutton constructed a cyclic theory of earth history and identified the processes that modern geomorphologists believe shape the primary landforms of our planet, the hills and valleys. By allowing long periods of time for landforms to evolve by uplift and erosion, Hutton did away with the need for catastrophic explanations of valleys and mountains. Before Hutton, massive earthquakes, volcanic eruptions, flooding by the sea and violent drainage of seawater from land were evoked to explain valleys and hills. Hutton developed what has become known as the principle of uniformitarianism: 'No powers to be employed that are not natural to the globe, no action to be admitted of except those of which we know the principle, and no extraordinary events to be alleged in order to explain a common appearance' (Hutton 1795, vol. 2: 547). In other words, only those ordinary processes of erosion and deposition that can be observed, and uplift that can be both measured and inferred, are permitted to explain landforms. Inferring uplift required geological proof however, and Hutton found it in unconformities — that is, where an erosional plane separates older rocks from overlying younger rocks (Davies 1968).

Hutton's cyclic earth history of uplift, erosion, deposition, lithification and uplift, has an important implication. Somewhere the rocks reduced by erosion to a landscape of gentler topography should have been buried by younger sediment that has been lithified. Therefore older rocks should be separated by an erosional plane (an unconformity) from younger rocks. Hutton found and illustrated several examples, but only after he had formulated his theory — an inversion of the way that most modern geomorphology proceeds from observation to generalisation and theory.

The eroded rocks were also seen to be affected by 'subterranean heat', an early and rather oblique reference to thermal metamorphism of rocks: that is, the physical and chemical alteration of rocks by heat. Uplift was also assisted by heat, and granite in Hutton's view was the result of molten rock rising into the crust where it eroded and crystallised into solid rock. This idea was radically different from earlier views, common in the 16th and 17th centuries, that the Earth's granitic topography was either formed at the time of the Creation and has not changed since or, like all other rocks, is the result of precipitation from fluids.

While Hutton was not entirely original (Davies 1968), he explained rocks and major landforms. He was also the first to opine that soil is produced by rotting of rock so that soil remains despite erosion. This, and his cyclic theory of earth history, overcame the 'denudation dilemma' whereby the Creation of a planet for humankind should not be subjected to decay and irreversible destruction by erosion.

Since Hutton, most geology and geomorphology has been focused on a fundamental understanding of the composition, processes of change and formation, and history of rocks and landforms. Both fields of study are strongly empirical, that is, observation is dominant over theorising. As physics and chemistry have evolved, they have provided insights and techniques for both geologists and geomorphologists, such as an understanding of the chemistry of weathering, the physics of sediment transport and deposition, and the differences in chemical composition of rocks, soils, waters, and sediments. And now there are techniques for determining the age of rocks, sediment and soil, thanks largely to developments in the chemistry and physics of radioactivity. Geographers too have had influence, by ensuring a strong interest is maintained in the spatial distribution of landforms, and by the integration of studies of land use with geomorphology and soil science.

2. Earth science becomes environmental earth science

Most geomorphology has been concerned with landform evolution and geometry over long time periods. W.M. Davis' geographical cycle, Penck's theory of denudation, and Strahler's statistical analyses of regularities are often cited as the foundational works. G.K. Gilbert in the United States of America has also made seminal observations of processes of landform development (Chorley et al. 1984).

The geographical cycle purported to explain the major features of the Earth's landmasses by postulating a neo-Huttonian sequence of uplift and denudation, and deposition of the products of erosion. But at that time there was little understanding of the varieties of uplift, their rates, and timing. And there was no allowance in Davis' ideas for interruptions to denudation by renewed uplift. In the last few decades geochemists, geophysicists and geologists have been able to estimate the timing of uplift by using radioactivity in rocks, and geomorphologists have married observation of landforms in uplifting and denuding landscapes with estimates of uplift rates and timing to provide a much more accurate picture of the dynamic surface of the Earth. Geologists, interested before this time in the deformation and metamorphosis of rocks in uplifted terrain, are now interested in the landforms as well. Computer modelling of uplift, mountain formation, rock formation, and erosion of mountains aided by the rainfall generated by the rising mountains, has demonstrated the intimate links between these phenomena (Kooi and Beaumont 1996). When coupled with the most pervasive and successful grand-scale theory of the earth sciences — plate tectonics — these studies are revolutionising our ideas of landforms. Plate tectonics explains that the crust of the Earth is made up of a number of semi-rigid plates that collide and slide past each other causing earthquakes and uplift. Where they diverge, mostly in the oceans, plumes of lava well up from below the ocean floor to form ridges. Where they diverge on land, huge rifts form, the best known being the East African Rift Valley. The floors of the oceans dive beneath the continents to form volcanic island arcs such as Japan, or force up landmasses by uplift and volcanic activity, such as in the Andes. And continents collide to produce massive mountain ranges such as the Himalayas (Skinner and Porter 1995).

These major insights of earth science provide in the case of the computer modelling of mountains an exemplar of convergence between geology and geomorphology. Furthermore, plate tectonics provides an explanatory context for aspects of modern environmental management, telling us for example where and why earthquakes, volcanoes and high soil erosion rates occur, and where and why they do not. But this work leaves the major environmental problems susceptible to the approaches of earth science largely untouched.

So what are the environmental problems that can be usefully tackled by the environmental earth scientist? The most significant globally are:

(a) soil erosion by water and landslides

(b) sediment transport through river catchments

(c) processes, rates, and causes of changes to the planform, cross-section, and sediment types in rivers and on coasts

(d) soil erosion by wind, and the movement of dunes, both in deserts and on coasts

(e) the movement of glaciers, and the production of sediment; volcanic debris behaviour

(f) soil formation rates, particularly in relation to soil erosion rates; lake pollution, ecology, and deposits

(g) impacts of earthquakes on landforms, and landslide generation

(h) the behaviour of estuaries, especially the transport and deposition of sediment and the creation and destruction of mudflats

(i) the changes wrought by past climate change as a guide to the possible effects of future climate changes, and the record of climate change recorded in ice, mineral sediments, and landforms

(j) the migration in groundwater of chemicals dangerous to living organisms, a topic especially pertinent to the management of the wastes left behind from mines where toxic compounds such as lead, uranium, and radium are involved.

This list of problems has two important characteristics. The first is that most of the changes involved occur over relatively short timescales relative to the time periods over which major landforms such as valleys change significantly. While timescales of a few millennia are valuable for explanation of current states, for example by comparing modern atmospheric warming with the record of past temperature recorded isotopically in ice, most environmental problems will be manifest following some human cause within decades to a few centuries. In some cases, a record of climate and attendant change to landforms and the biota over several hundred thousand years can help explain the functioning of the Earth system. By comparison, thousands to millions of years are required for soil formation and major valley formation, respectively.

The second characteristic is that the list has no evident reference to the fundamental principles that geologists and geomorphologists have identified over two hundred years. This lack of connection between earth science and environmental earth science is seen in textbooks where most of the book is devoted to earth science and the last chapter includes people and environmental issues, usually by using case studies rather than reference to basic principles. But it is straightforward to relate earth science and environmental earth science, as shown below.

3. Approaches and key concepts of environmental earth science

The environmental earth scientist is concerned with two cases of environment–human interaction: human agency as a cause of change to soil, water and landforms; and human behaviour in the face of these changes and phenomena, including natural hazards such as earthquakes and storm surges. An example of human agency causing deleterious change is where land use in a river catchment increases erosion of soil and therefore the amount of sediment transported in a river. This additional load can destroy aquatic habitat by infilling pools in riverbeds, and can infill reservoirs thereby shortening their useful lifespan. When people build on floodplains, or within the zone of storm surges produced by tropical cyclones (hurricanes), they increase their risk and exposure to natural hazards. This is an example of the second case.

Both of these approaches involve process, history and human behaviour. In the first case, land use change and the erosional response, and then the movement of sediment into a river (to be partly deposited in a floodplain and then removed from this site of temporary deposition as the river moves laterally into the material it has previously left behind) can take decades to centuries to reach a stage where change is slow once again. In the second case, the frequency of large and damaging floods or storm surges is often low enough for there to be little or no memory or documentation. The techniques of historical geology can be used to determine when in the past such floods and storm surges have occurred, and therefore their frequency.

There are therefore three types of investigation that are most important in environmental earth science. The first is the study of processes and behaviour of earth materials by direct observation either in nature (in the field) or in the laboratory. Examples include measurement of the quantities of sediment transported in a river in relation to the velocity and quantity of water in the river, and the forces applied to the sediment by the moving water; rates of movement of dangerous chemicals in groundwater, and the form and alteration of these chemicals as they are transported; and changes to the resistance to landsliding of hillslope soils as trees are removed.

The second type of investigation uses historical studies of erosion, transport, and deposition of sediments to deduce trends and frequencies. In rare cases, instrumental observations of river flows or earthquakes are long enough to reconstruct trends and frequencies. But usually these characteristics need to be reconstructed from deposited sediments. In the case of floods, sand layers in floodplains record large flows. By dating these sand layers, the frequency and change of frequency of floods can be estimated over a sufficiently long period of time to give confidence in the result. The same approach can be used in lakes to estimate the frequency of inflow events, the effect on lake biology of pollution, and the effect of pollution control programs when the study of layered deposits in lake floors is combined with age estimates and chemistry.

The third type of investigation is the study of the variation of process and change over areas of differing size. The movement of wind on a dune produces ripples that are organised over areas of a few tens of square centimetres. Dunes are larger structures, forming features tens of metres wide and kilometres long, but formed in the same wind field and by the same inter-particle movement as ripples. These patterns emerge at different spatial scales, and can be altered by a change of wind or overgrazing of the vegetation that minimises sand movement. Similarly, increased soil erosion by water in response to replacement of native forest by agricultural fields can lag behind the land use change by 10 or so years as soil organic matter decays and the soil consequently becomes less able to resist erosion by rainfall. In the river, the passage and deposition of sediment released from land use change upstream can occur many decades after soil erosion on cleared and cultivated hillslopes has occurred. Moreover, the increase (over the pre-agricultural state) of the amount transported or deposited varies along the river, becoming less significant further downstream.

The concept of uniformitarianism runs through all three kinds of investigation. As we have already seen, Hutton did away with any need for catastrophic change to explain either landforms or rocks, by claiming that all former continental masses were of the same materials as the modern continents, and that the types of processes and their speed of action had not changed through time. That is, the slow processes of uplift, denudation, deposition and lithification seen today have always occurred. This is 'substantive uniformitarianism', to which has been added since Hutton's time the notion that chemical laws do not change with time.

By the 1830s Hutton's cyclic earth history had been accepted at least in Britain, but there was disagreement about the nature of uplift. Some geologists followed Hutton, seeing the Earth as forever moving, and uplift as a perpetual and extremely slow process. Other geologists believed that uplift was achieved by violent and catastrophic earthquakes (seismic events). Charles Lyell, one of the great 19th century geologists, argued the uniformitarian view against catastrophists in both Britain and Europe (Davies 1968).

More and more evidence of very large but infrequent events in earth history has come to light since Lyell's time — large meteors striking the Earth, extinctions of animals and plants, and huge floods such as the one that drained from glacial Lake Agassiz in North America to produce the 'Channelled Scabland' (Bretz 1928). Neo-catastrophism has grown in popularity as evidence has accumulated.

4. Magnitude and frequency analysis

A key analysis by Wolman and Miller (1960) has largely resolved the disagreement between substantive uniformitarianism and neo-catastrophism, to produce what can be called weak uniformitarianism in which no process is permitted that is 'supernatural', but the rare and extremely large (catastrophic) event is accepted. Wolman and Miller used a simple graph to illustrate their idea (see figure 8.1). Curve 'a' shows how the rate of transport of (say) sediment in a river or on a dune increases with the size or magnitude of the applied stress. That is, as velocity of water or wind increases, so does the rate of movement of sediment. Notice that the rate of increase of sediment transport increases faster with the applied stress, a characteristic of all forms of sediment transport. The curve can be mathematically described by a power law.

The applied stress is provided by many individual flow events, windstorms, or wave sets. The magnitude of these stresses is most readily summarised as a frequency distribution (curve 'b' in figure 8.1), which begins at zero, rises to a peak and then trails off at high applied stress. That is, most transporting events are of low to moderate magnitude, and very high magnitude events are very rare.

The maximum amount of sediment transported, expressed as maximum total work in figure 8.1, is estimated by multiplying curves 'a' and 'b'. The maximum total work occurs at an applied stress slightly higher than the peak of the frequency distribution for transporting events because of the curved form of 'a'. The curve of maximum total work is also displaced to the right because there is a threshold value for the beginning of curve 'a', shown by t_1 in figure 8.1. The threshold indicates that a certain value of applied stress (e.g.

water velocity or windspeed) must be exceeded before sediment will move.

Because this concept was constructed from data collected in the humid, low relief, soil-covered landscape of the eastern USA, the peak of maximum total work was found to be in the middle range of applied stress. In environments where the threshold of movement is high, because soils are cemented by iron or silicon for example, only rare high magnitude events will have any impact on landforms. So the peak of total maximum work is shifted to high applied stresses.

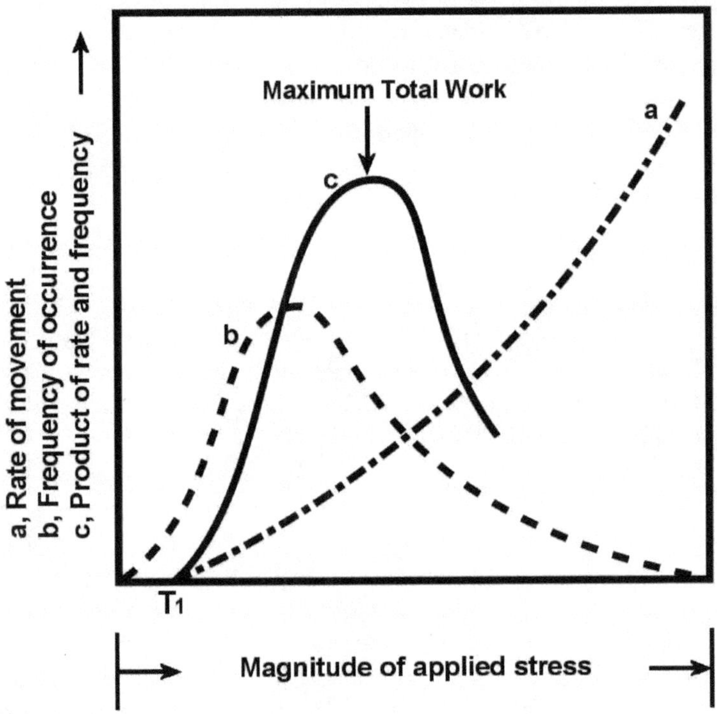

Figure 8.1. The magnitude-frequency spectrum. The rate of movement of sediment increases along curve 'a' as flow velocity (applied stress) in, for example, a river rises. The frequency of each value of applied stress is given by curve 'b'. The amount of sediment moved, and the applied stress at which most sediment (work) occurs, is given by multiplying together curves 'a' and 'b' to produce 'c'.

5. People and the earth

As we have seen, environmental earth science mainly focuses on the processes, soils and landforms that respond relatively quickly to human activities, and on processes that occur relatively frequently and constitute a hazard for both human populations and ecosystems. The landforms that respond most quickly are made of sediment and soil, rather than rock, located usually in the energetically most active parts of the landscape such as along rivers. Because of this focus the environmental earth scientist pays particular attention to relatively short timescales, usually a few millennia at most. From sequences of deposits, the frequency of rare but large events can be deduced to complement the shorter records based on monitoring. Trends can also be established.

The magnitude–frequency analysis of figure 8.1 then allows planning and risk assessment of natural hazards, as well as estimates of the probability of different magnitudes of erosion and deposition given long enough records from which curve 'b' can be accurately estimated. Curve 'a' is derived from laboratory and field measurements, but its upper end is poorly defined because measuring transport rates at high applied stresses in rivers, for example, is hazardous and only rarely done.

Examples of integration between environmental earth science and human concerns are numerous, but groundwater is especially pertinent. Sustainable groundwater use is easy to define: the rate of extraction must be less than or equal to its replenishment, and it must not be polluted to a level beyond which it either cannot be used or will harm the life that depends upon it. The key concepts here are the water balance and understanding how pollutants reach groundwater and can be kept separate from it.

Measuring water balance requires estimates of recharge (replenishment rate) by infiltration of rainfall through soils and into groundwater, the rate of lateral inflow in the subsurface from upslope, the rate of loss into rivers and the ocean by seepage, and the rate of extraction both by vegetation through roots and by human activities. Measurement of the components of the water balance can be done by gauging and chemical methods (Mazoor 1997).

Determining the causes and impacts of groundwater pollution can be much more difficult, as exemplified by arsenic poisoning in both Bangladesh and the Indian state of West Bengal (Rahman et al. 2003). An estimated 20 million people here are drinking water poisoned by naturally occurring arsenic, causing skin pigmentation, warts, diarrhoea, ulcers, and cancer. There are at least two possible causes of the arsenic poisoning. Beneath the floodplains and rivers of the region lie sediments deposited in mangrove swamps and in rivers near the sea. Large amounts of arsenic, derived from the Himalayas, were included in various forms of the mineral pyrite in these sediments, and then buried. Where tubewells were sunk to provide clean drinking and irrigation water, oxygen reached these pyrites and released arsenic. The second possible cause relates to the unusual release of arsenic from the surfaces of clay particles and from compounds of iron, aluminium and manganese in the sediments from which drinking and irrigation water is taken.

Environmental earth scientists are actively discovering the causes of the poisoning, locating layers of sediment with water free of arsenic, and working with engineers and health professionals to treat water so that it can be used. This example also highlights the difficulties of prediction. While arsenic poisoning has been reported in Argentina, Finland, Mexico, Nepal, USA, and Vietnam, nothing on the scale of the problem in Bangladesh was imagined when groundwater was tapped.

In summary, environmental earth science is the study of the impact of people's activities on earth surface (and near-surface), water bodies, processes and landforms, and the frequency and magnitude of processes that affect people and/or ecosystems. These studies involve an understanding and observations of key processes, a history of processes and their outcomes, and an investigation of variation in rates and rates of change of processes, water bodies and landforms over areas of varying size and location. Prediction remains a key challenge.

These basic elements of environmental earth science are now further illustrated by examining soil erosion and transport of sediment in rivers.

6. From hillslopes to the sea

The conversion of river catchments from native vegetation to pasture or cultivated land almost always increases erosion of soil by sheet erosion of topsoils and by gullying. The removal of the vegetation along river margins (the 'riparian zone') results in enhanced river-bank erosion, which is sometimes exacerbated by greater riverflows caused by increased runoff following conversion of catchment vegetation. Higher than 'natural' loads of sediment in rivers can have serious impacts, both economic and ecological, as described above. Higher than 'natural' soil erosion rates deplete an essentially non-renewable resource, with serious economic and social consequences.

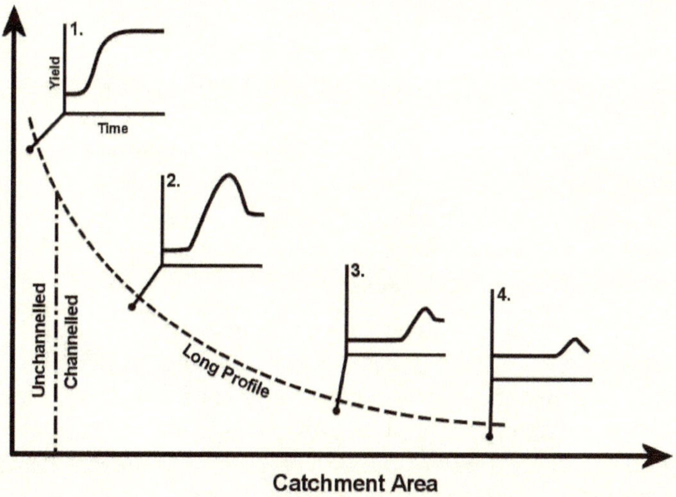

Figure 8.2. Responses (expressed as yield) to a change (such as clearing) in a river catchment, at various points along the river. At point 1, land clearing leads to a sustained increase in loss (yield) of soil from hillslopes. At point 2, yield measured in a river rises as erosion from hillslopes increases, channels enlarge, and gullies develop. As gullies and channels reach their maximum dimensions, but hillslopes continue to erode, yield declines. The same occurs at points 3 and 4, but the beginning of the yield increase starts later because time is required for sediment to be moved from upstream to these points in the river. The yield increase is also smaller because some sediment from upstream is lost by deposition on floodplains.

One way of integrating the change of erosion rate and river sediment transport rate in a catchment is shown in figure 8.2 (from Wasson 1998). A change of vegetation cover in the uplands, that is in the area where channels do not occur, is responded to by a dramatic increase of soil erosion rate by sheet erosion of topsoils (position 1). There is a lag between vegetation cover change (assumed to be at time zero on the horizontal axes of the graphs in figure 8.2) and increased soil erosion because, as noted earlier, soil strength and resistance to erosive rainfall takes some time to decay. Within the channel network, the response of river sediment transport rate lags behind the increased soil erosion rate in the uplands because of the time required for the sediment to reach position 2. If gully and channel erosion also contribute to the river sediment load, the transport rate curve at position 3 will fall once a peak has been reached because gully extension and channel widening slow to reach a new quasi-equilibrium stage. At position 3 the response is even more delayed and the peak smaller because much of the sediment is deposited on floodplains between positions 2 and 3. This explanation also applies to position 4.

The graphs in figure 8.2 are plausible but largely hypothetical. To document such graphs along a river requires the use of a wide variety of techniques and data, as follows:

Position 1 Record of land use and land cover change – measured rates of soil erosion of topsoils and gullying from before land use change; estimates from field survey of soil loss relative to a datum such as the previous thickness of topsoil or the pre-gullying form of valley bottoms; or estimates based upon mathematical models of erosion processes calibrated for local conditions.

Positions 2, 3, 4 Measured loads of sediment in the river before land use change, or modelled estimates of load based on calculations of delivery of sediment to the river from all sources of erosion, corrected for storage on floodplains in particular, and estimates of the river's transport capacity through time from its flow characteristics. It is also possible to identify a peak of sediment transport following a land use change from deposits stranded in the mouths of tributary streams, or from enhanced sedimentation on floodplains. Dating of sediments is critical at such sites, and the recent development of luminescence dating provides an appropriate technique.

Among the most useful information that can be supplied to a catchment manager is an estimate of the source of sediment infilling, for example, a reservoir. Once the source is known, catchment remediation can be efficiently designed. Sources can be thought of as topsoil and topsoil plus subsoil, the former eroded by sheet flow across soil surfaces and the latter by gullying and channel erosion. Direct measurements of these processes of erosion can be made of the quantities of material coming from each source type and from different sub-catchments. To this can be added an accounting of deposition on floodplains for example, and of the loss (yield) from a catchment and its impact by sedimentation on the lifetime of a reservoir (Wasson 1998).

Topsoil accumulates radioactive materials from the atmosphere, such as a naturally occurring lead isotope Pb-210 and a product of nuclear weapons testing, caesium Cs-137. When topsoils are eroded these isotopes are found in high concentrations in river sediment. When subsoils are eroded by gullying, these isotopes are found in low concentrations, because they only penetrate up to ~20 cm below the ground surface (Wasson 1998).

Different sub-catchments of a river basin often contain different rock types with differing chemistries and mineralogies. These differences can be used to determine the proportionate contribution to, for example, the sediment in a reservoir from different sub-catchments.

A relatively new technique now allows estimation of soil formation rates and sediment transport rates, and when combined with traditional measurements of sediment loads in rivers and erosion rates estimated from troughs installed on hillslopes, powerful insights into the dynamics of sediment production and transport can be gained. Hewawasam et al. (2003) have studied the Upper Mahaweli catchment in Sri Lanka. Fine sediment loads in rivers show that the loss of soil is between 130 and 2100 tonnes/km^2/year. Measurements of eroded soil trapped in troughs on small plots in agricultural land show a rate between 100 and 7000 t/km^2/year. Natural, that is pre-agriculture, rates of sediment generation are only 13 to 30 t/km^2/year. These rates are estimated from the amount of a rare isotope of beryllium, Be-10, in soils and sediment. This isotope is produced in the upper atmosphere by cosmic rays and is deposited in soils. The amount of Be-10 in soil and rock is inversely proportional to the total erosion rate.

These results indicate a 10 to 100-fold increase in soil erosion as a result of agriculture. Soils undisturbed by cultivation reside on hillslopes for between 60 000 and 150 000 years. Soils under agriculture will reside on hillslopes for as little as 250 years. This is clearly an unsustainable situation where soils are being removed far quicker than they are forming.

7. Conclusions

Environmental earth science is intimately involved with human affairs, through people as an agent of change and because people are affected by changes to the natural world. Some of these changes are catastrophic from a human perspective and are called hazards. But they are hazards because of human behaviour and choices, not simply because of the occurrence of (for example) large earthquakes and floods. Therefore, to make a real contribution to natural resource and environmental policy and management, environmental earth scientists must engage with their colleagues in the social sciences (including economics) and the humanities.

Environmental earth science nonetheless provides concepts, frameworks and techniques that form a rigorous basis for connection to other fields of knowledge. Environmental earth science is also in its own right a useful field of study, because it can be connected directly to natural resource and environmental management. Advice to catchment managers can come from studies of (for example) sediment sources. So environmental earth science can be of practical value both alone and with other fields of study.

8. Further reading

A varied and very useful collection of papers on the philosophy of science, the origins and nature of geomorphology and its applications is provided by Rhoads and Thorn (1996). For an important paper in applied geomorphology in this collection, see W.L. Graf (1996). The first collected papers in applied geomorphology are those edited by Coates (1972, 1976).

9 HYDROLOGY

Ian White

1. The science and engineering of water

Water is vital to all life on Earth. It is also a major economic resource and has deep cultural, social and religious significance in many societies (see chapter 3). Tensions between using water to sustain populations, maintain their environments and ensure economic development are increasing. Population growth over the next century will affect the availability and quality of water resources and threaten the health of ecosystems. Some estimates suggest that by 2050, two-thirds of the world's people will live in water-stressed countries, particularly in India and large parts of Africa and China (Falkenmark 1997). Already, millions of people die each year from waterborne diseases and, in some areas, women walk up to 20 kilometres each day to fetch tiny amounts of water of dubious quality for their households. Paradoxically, engineering solutions for some of these problems have also had adverse impacts on some societies, their economies and downstream environments.

Humankind has an increasing capacity to alter the water cycle both locally and globally. The challenges of water scarcity are compounded by continued contamination of rivers, lakes, groundwaters, estuaries and nearshore environments, particularly through the growth of mega-cities, and by our limited knowledge of the quantities, flows and conditions of water resources and the impact of global change on them (Gleick 1993). Hydrology — the geoscience and engineering of water — seeks to address these challenges and knowledge gaps, but it alone cannot solve the complex water resource problems facing societies. It provides much of the information required for equitable and rational water resource policy and use, but it is up to societies to make informed choices about use, management and allocation of their water resources (Bredehoeft 1997).

Definitions

Hydrology is a component of the broader fields of environmental earth science (chapter 8), of environmental mechanics (Raats et al. 2002), the applied mathematics of the environment, and of civil engineering.[8] At its broadest, hydrology is the science that treats the waters of Earth, their occurrence, circulation, and distribution, their chemical, physical and biological properties, and their reaction with the environment, including their relation to living things (Maidment 1992). Hydrology considers the interactions between the hydrosphere, lithosphere, biosphere (including humankind) and atmosphere and involves, at its core, the intimate connection between the Earth's water cycle, the spatial and temporal distributions and variations of water, and its energy balance.

Initially, hydrology was principally concerned with the physical properties of water in the hydrological cycle encompassing precipitation, evaporation, runoff, streamflow, infiltration (the process of water entry into the soil), soil and groundwater flow. As our understanding matured, this expanded to include the transport of suspended materials and dissolved chemical species, as well as the biological properties of water and its interaction with the environment and living beings. The disciplines that contribute to hydrology include mathematics, statistics, physics, chemistry, biology, microbiology, botany, plant physiology, and the earth sciences of fluid mechanics, geology, geophysics, geochemistry, geomorphology, physical geography, soil science, atmospheric science, ocean science and glaciology.

The study of hydrology evolved from two separate approaches. The geographical earth sciences approach (see also chapters 8 and 11) seeks to explain the processes involved in the water cycle and the spatial relations between these processes. The engineering approach attempts to find ways of controlling water flows, particularly during extreme events, and to alter the distribution of water in order to service and protect human settlements and their food and fibre production systems. While these disciplines overlap enormously, the geographical treatment has tended to be more descriptive while the engineering approach is more mathematical. In Europe, some distinction is made between scientific hydrology, with an emphasis on description and explanation, and applied or engineering hydrology, with a focus on solutions to operational problems.

The processes involved in the hydrological cycle can vary spatially from micrometres (water films in dry clay soils) to thousands of kilometres (the global scale) and over time from seconds (atmospheric turbulence) to millions of years (deep groundwater residence time) (Dooge 1997). These enormous variations in scale pose significant difficulties for any systematic description of water. Often, the importance of hydrological processes can also vary with scale. For example, change in soil water storage, so important to plant growth at daily or weekly timescales, is negligible globally at yearly timescales. In addition, processes can often be highly non-linear. Some processes vary enormously with modest changes[9] and may not be directly reversible. All these reasons complicate the description and modelling of hydrological processes.

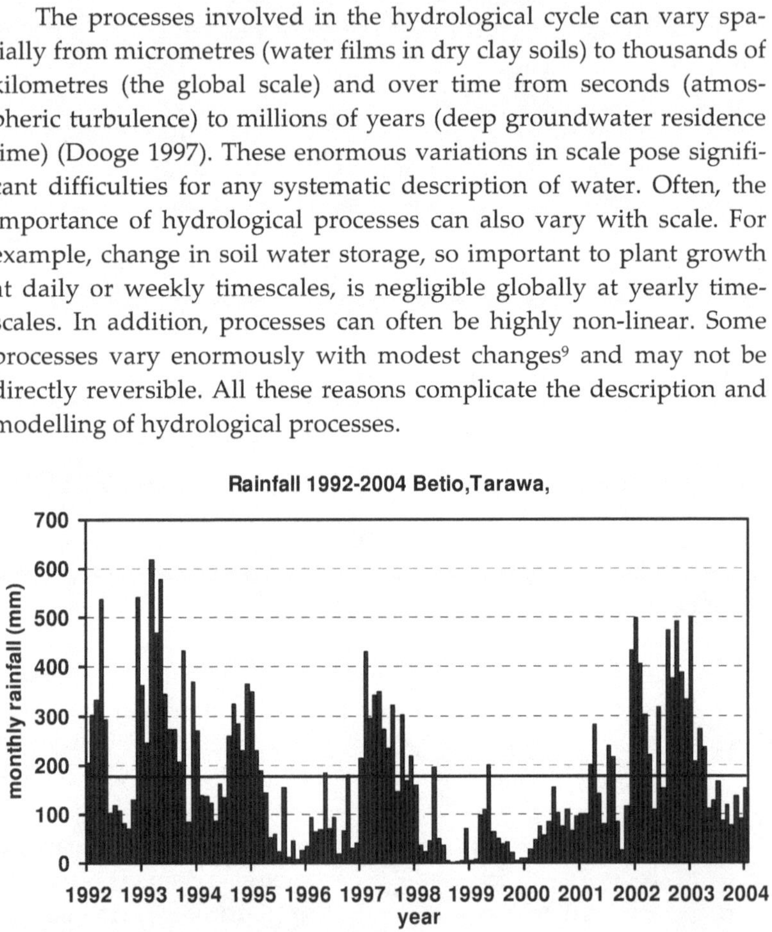

Figure 9.1. The extreme variability of recent monthly rainfall (coefficient of variability is 82%) for the central Pacific atoll of Tarawa, Republic of Kiribati. Rainfall in Tarawa is highly correlated with El Niño events as the Pacific warm pool sweeps back and forth across the central Pacific. The line shows mean monthly rainfall for the period. The variation is skewed towards higher values.

The systematic collection and analysis of measurements of components of the water cycle, such as rainfall, streamflows (or their surrogate stream heights), solar radiation, wind speed and humidity, pan evaporation, soil water content, groundwater elevation, sediment loads, water chemistry, biology and microbiology, are fundamentally important in hydrology. These data sets, often described as point measurements because they are measured over relatively small areas,[10] both inform and provide tests for hydrological theories, models or laws of dynamic processes. Importantly, they provide evidence for natural and human-induced changes. Because the collection of data is expensive, governments are increasingly reluctant to support extensive long-term hydrological monitoring.

Point hydrological data sets are often characterised by extreme variability in space and in time (see figure 9.1) and a range of statistical processes have been developed specifically to handle this variability. In many cases the variability is not normally distributed about the mean but can be skewed towards higher (as in figure 9.1) or lower values.

Origins

The first settlements developed on stream banks and, since water is both a vital necessity and a natural hazard, the quest to understand and even control water stretches back to the Neanderthals. Evidence for the construction of canals, levees and dams, underground conduits and wells in the Tigris and Euphrates rivers in Mesopotamia, the Indus in Pakistan, the Nile in Egypt and the Hwang Ho in China date back to as early as 5000–6000 years ago (before present — BP).

River flow in the Nile was monitored by 3800 BP. The timing of floods in the Nile was so accurate it provided the most precise estimate of the length of a year until the 18th century. Rainfall measurement commenced in India around 2300 BP (Dingman 2002). Written records demonstrate that appreciation of a rudimentary form of the global water cycle existed by at least 3000 BP.

> All streams run into the sea, yet the sea never overflows; back to the place from which the streams run they return to run again (Ecclesiastes ch 1 v 7, *New English Bible*).

The mechanism for the return of water from sea to land remained a mystery even to the early Greek philosophers. By about 2350 BP Aristotle appears to have deduced the mechanics of precipitation and three hundred years later the Roman, Vituvius, inferred the origin of springs. The Romans were outstanding hydraulic engineers. Some of their aqueducts are still in use today, 2000 years after they were built. They did not fully understand why they worked as they did. When flow in aqueducts decreased due to increases in channel cross-sectional area, they believed it was due to water theft, not natural causes. We know now that this decrease is a simple consequence of conservation of mass (see below).

Greek philosophical thought dominated European thinking about water until around 1500 when, through field observations, both Leonardo da Vinci in Italy and Bernard Palissy in France reasoned that river flow is driven by precipitation. This marked the beginning of scientific hydrology. In the late 1600s Pierre Perrault and Edmé Mariotte in France verified quantitatively that rainfall drives streamflow. In about 1700 the English astronomer Edmund Halley was able to estimate the quantities of water involved in the ocean–atmosphere–river cycle in the Mediterranean Sea (Dingman 2002).

The term hydrology, associated with a rudimentary form of its current meaning, appeared by about 1750. The 18th century saw the blossoming of the application of mathematics to fluid mechanics by renowned European mathematicians and natural philosophers whose names, such as Pitot and Bernoulli, are still associated with aspects of fluid flow. By the close of the century, the brilliant English physicist and chemist John Dalton had fully described the nature of evaporation and the global hydrological cycle. Work on fluvial erosion of valleys was initiated by the beginning of the 19th century. While rainfall has been monitored in Korea since the early 1700s, networks for routine measurement of precipitation were not established in Europe, the United States and India until 1820.

Groundwater flow continued to remain a mystery until the work of the French engineer Henry Darcy in 1856. Darcy had set himself the task of providing pure water for his home town, Dijon, in the 1830s. His work on sand filtration of water provided both the foundation for modern urban water treatment and the phenomenological law of fluid flow through porous materials. Darcy refused any form of remuneration or reimbursement for the twelve to fifteen years of work he devoted to transforming the quality of Dijon's and thereafter the world's urban water supplies (Philip 1995).

The 19th century saw great advances in fluid mechanics. The names of the mathematicians, physicists and engineers such as Poiseuille, Stokes, Reynolds and Manning are associated with flow principles and laws discovered over this period. The first English manual of hydrology, Nathaniel Beardmore's *Manual of hydrology*, was published in London in 1862.[11] Most early works concentrated on the relationship between streamflow and rainfall in order to estimate flood flows for the design of bridges, levees and dams. This close association of hydrology with civil engineering has continued to the present.

The first formal hydrological societies were formed early in the 20th century and this marked the wider recognition of hydrology as a science. The 20th century saw the rapid evolution of hydrology. Developments included: statistical tools to describe and predict hydrological processes; theoretical and experimental studies of groundwater flow, stream hydraulics and sediment transport, infiltration, unsaturated soil water flow, evapotranspiration,[12] runoff generation; quantitative geomorphology; the application of systems and time- and spatial-series analysis (see chapter 10); global-scale hydrology; and the understanding and modelling of water quality and algal blooms. These developments have not just involved the application of extant science to hydrology. In some cases, new areas of inquiry such as dam theory in statistics[13] have resulted from the study of reservoir water storage problems.

Computer technology has brought with it a rapid growth in hydrological research. Some have lamented the increasing reliance on the output of computer models as reality rather than actual measurements (Philip 1998). While pioneering and surprisingly accurate work on the broad features of global-scale hydrology was initiated in Russia in the early 1900s, computers have permitted significant advances in the study of global-scale dynamics (Budyko 1974).

Recent developments

Many recent developments in hydrology involve interactions between several disciplines. Recognition of the relationship between catchment topography, saturated areas and runoff has improved our understanding of catchment processes. Convergent surface topography channels subsurface flow to lower lying, flatter areas, which become saturated. Runoff is thereby related to the size of the saturated areas surrounding streams (see, for example, O'Loughlin 1981). Appreciation of the links between the great circulation currents in the ocean, particularly in the Pacific, and droughts and floods is still evolving. However, the relation between the difference in surface atmospheric pressures in Tahiti and Darwin represented by the Southern Oscillation Index (SOI) and severe droughts or floods in many continents is sufficiently well known to be quoted on evening television weather reports.

The use of proxy data such as the stratigraphy and geochemistry of sediment, coral and ice cores to deduce palaeoclimates is greatly extending our limited meteorological records and our understanding of extreme events (see, for example, Fairbanks et al. 1997). In slowly growing corals or accreting ice or sediment layers, the depth profile of ratios of different chemical species or isotopes[14] of elements can be used to infer past temperature and rainfall regimes, in some cases increasing the measurable records to 450 000 years.

The application of environmental isotopes has enabled better identification of processes and estimation of rates of transfer. For example, the use of the radioactive caesium isotope, caesium-137, a fallout product from atmospheric atomic bomb testing, can be used to infer sources of top soil and rates of soil erosion from sediments deposited in lakes and dams. As well, enrichment in water samples of the heavy isotopes of water, deuterium (hydrogen-2) and oxygen-18 can be used to trace past evaporation histories.

The development of remote sensing by airborne or space platforms has allowed, for the first time, the monitoring of hydrological properties at larger scales appropriate for the management of resources. Land surface temperatures and reflectance in different parts of the electromagnetic spectrum enable the estimation of soil water content and the prediction of crop yields. As well, reflected light from water bodies permits the early detection of potentially hazardous algal blooms in remote locations.

Better understanding of the microbiology and virology of water systems is providing both increased appreciation of the risks involved in water supplies and new ways to treat water. In one application, polluted city stormwater is deliberately recharged into groundwater aquifers that act as microbiological reactors. Bacteria in the aquifer use the pollutants as food sources, thereby purifying the water which, free from evaporation losses, is then available for extraction and reuse.

Ecohydrology is a relatively recent advance that deals with the fundamental relation between ecology (see chapter 7) and water (Eagleson 2002). Ecohydrology seeks to understand and predict the distribution and behaviour of plant, aquatic and animal species from the prevailing water, nutrient and temperature regimes in an area. For example, it was recently recognised that dust storms and massive forest fires emit large quantities of iron into the atmosphere. When deposited in oceans, the iron is a nutrient that can trigger algal blooms and damage coral reefs in areas far removed from the dust source or fires.

Finally, recent work has attempted to combine rules of behaviour for social, cultural and economic factors with hydrological models in order to provide more equitable avenues for water use and sharing. These are mostly based of the use of multi agent-based simulations that try to capture the dynamics of behaviour and decision-making over resource use. See, for example, Bousquet et al. (2001).

2. Concepts and fundamental principles

A fundamental principle in all earth sciences is the assumption that processes taking place and observable today are the same as those that occurred in the past. We assume that processes continuing since the establishment of the hydrological cycle have shaped the present characteristics of water on Earth.

The hydrological cycle

The hydrological cycle and its interaction with Earth's energy balance are two important fundamental concepts in hydrology. Incoming radiation from the sun is partitioned at the Earth's surface. Some is reflected (sensible heat), some is used to evaporate water (latent heat), and a much smaller fraction is transmitted into the earth. The surface partitioning of solar radiation on land is critically dependent on the availability of water at and close to the surface.

Evaporated water is transported away from the surface as water vapour by atmospheric circulation. When cooled, and provided there are sufficient nucleation particles[15] in the atmosphere, water vapour condenses to liquid or solid water and eventually falls as precipitation (rain, snow, hail, sleet, drizzle, dew). At the Earth's surface, water can be intercepted by plant surfaces, can become runoff to streams, lakes and the ocean or can infiltrate into the soil. Most of this surface and near-surface water is available again for direct evaporation or for transpiration through plants. The fraction of precipitation converted to evaporation over long timescales (typically one year) provides a useful measure of the aridity of a location through the ratio of either precipitation to potential evaporation[16] or the ratio of actual evaporation to potential evaporation. When these ratios are substantially less than one, a location is arid.

A small portion of water that infiltrates into the soil eventually reaches the groundwater, most of which eventually discharges to rivers, lakes, seeps and springs or into the oceans. Solar radiation is the energy source relentlessly driving the hydrological cycle. Water movement is driven mainly by gravity, but in soils and plants the attractive forces between water and solid surfaces ('capillarity') also drive flow.

Conservation of mass and energy

The conservation of mass and energy, the water, material and energy balances are equally important fundamental concepts in hydrology. For a specified region such as a concrete tank (figure 9.2), a catchment, the ground surface, a dam or a groundwater aquifer, and a specified time period, the conservation equation for mass and energy is simply:

Amount of inputs = amount of outputs + change in storage + quantity transformed[17]

The conservation equation is a fundamental principle, not just for hydrology and water management, but for environmental management in general. We ignore it at our peril and that of future generations. In some catchments in Australia, water licences for irrigation (an output) can exceed average annual rainfall (an input) by up to 600%.

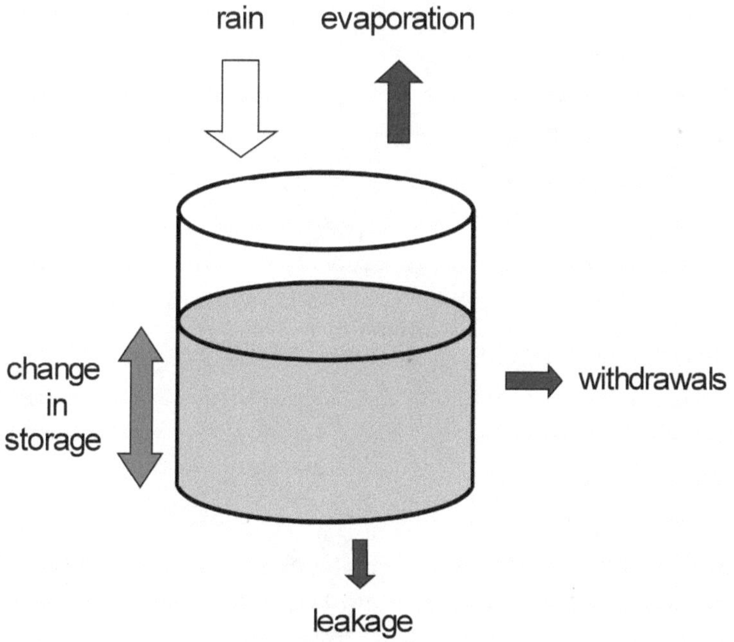

Figure 9.2. The water balance for a simple open concrete tank which is fed by rain (input) and from which water evaporates, is withdrawn for use and leaks out (outputs). If the input exceeds the output, then the water level in the tank rises as the tank stores more water. If the outputs are more than the inputs, the water level falls and water storage decreases.

Flow laws

Further important concepts in hydrology concern the relationships between rates of movement of water or materials, the driving forces that cause them to flow and the forces that oppose flow. These relationships are called flow laws. In principle, the movement of water vapour in the atmosphere, the planetary boundary layer, within plant canopies and the stomata of leaves; liquid water flow in plants, across the surface and through soil, in groundwater and in streams; the macroscopic circulation of lakes and oceans; the inexorable creep of glaciers; the transport of sediment; the dispersion of soluble and suspended materials in rivers, soil and groundwater as well as the mass movement of wet soil and rocks in landslips are all predictable from flow laws.

One of the simplest flow laws is the relation (law) for flow through porous materials (soils, aquifers) deduced by Darcy in his experiments with sand filters, published in 1856 (Philip 1995). For low velocity flows, Darcy's law states that the quantity of water flowing per unit cross-sectional area of porous material per unit time (a macroscopic flow velocity) is proportional to the water pressure head difference across the porous material (the hydraulic gradient). In this law, the driving force is the difference in water pressure across the material. The resistance to flow is lumped into the proportionality parameter, called the hydraulic conductivity. This parameter incorporates the microscopic properties of the porous material and the macroscopic properties of the water. The magnitude of the hydraulic conductivity depends on the square of the characteristic mean pore size of water-filled pores in the porous material and the viscosity and density of water in the porous material. When the amount of water in the porous material varies, such as when soil dries, the size of the water-filled pores decreases so that flow through the porous material is strongly dependent on water content.

When Darcy's law is combined with the water balance equation a flow equation is produced that describes quite generally water flow in soils, rocks, concrete, industrial slurries and filters. The law, however, is *phenomenological* (that is, specific to particular cases, not a universal law), because it requires that the spatial dependence of hydraulic conductivity and water pressure head on water content of the porous material be known (measured) and the distribution of water or pressure head be known at least at one time and one location.

Other basic laws

The laws deduced for the flow of materials and energy in hydrology rest on the conservation of mass and energy and on the laws of classical science such as Newton's laws of motion, Fick's law for the diffusion of dissolved substances and the laws of thermodynamics (heat and energy). Because of this, the ultimate validity of flow theory in hydrology rests on the validity of the application of these laws to natural highly variable systems.

3. Hydrological methods

The art of the falsifiable

There is no fundamental theorem or universal law from which all hydrological behaviour can be deduced. As an observational earth science, the methods used in hydrology rest on the methods and assumptions of its composite scientific and engineering disciplines. At their simplest, these include the assumption that there are measurable patterns of behaviour of water in natural systems, which, given the same conditions, are repeatable or follow what we call laws of behaviour. Once we have understood and articulated that law or advanced a model (conceptual or mathematical) or a hypothesis we may be able to predict the behaviour of water under quite different conditions. Science, being the art of the falsifiable, requires the testing of hypotheses, laws or models against actual behaviour.

Measurements and observations

Measurements and observations play a crucial role in hydrology. Laboratory experiments have played and continue to play a key role in understanding the basic processes in hydrology. In the field, if measurements of actual behaviour under changed conditions differ from our model predictions or hypotheses, then there are four possibilities: our model is wrong; our application of the model is faulty; the measurements are flawed or both model and measurements are in error. Given the large areas of the landscape that hydrologists need to understand, point measurements of hydrological components pose problems. We frequently assume that point measurements, such as rainfall or evaporation, are representative of much larger areas. Fortunately, statistical methods have been developed to test the representativeness of point measurements for longer timescales (see chapter 10). In addition, point measurements of hydrological properties are subject to significant errors. For example, rainfall, the most basic of hydrological measurements, can seldom be measured to an accuracy of 10%. As well, the enormous spatial and temporal variability of many hydrological processes challenge the application of scientific method to hydrology (Philip 1975).

4. Applications of hydrology

The contributions of hydrology, and particularly engineering hydrology, to human health and wellbeing have been spectacular (see chapter 5). Most aspects of the hydrological cycle from scales spanning the global to the microscopic are now understood. Indeed, engineering hydrology has changed the character of the hydrological cycle in many areas by altering the stocks and flows of water. Cities are able to survive droughts because of storage dams. Perhaps the most important contributions of hydrology have been in the areas of public health and safety. Darcy's experiments on sand filters and his application to the water supply and treatment for Dijon brought about one of the most significant improvements in public health. In 1920, Darcy's nephew described the condition of the waters of the Suzon, a small stream running through Dijon, before Darcy's time as follows:

> The Suzon ... served over a length of 1300 metres as the main sewer for wastes of every kind, for dwellings and adjoining industrial establishments (such as the tripe makers of the rue du Bourg); ...it was never cleaned; ... it was never flushed out, and ... above all during hot weather, the town was poisoned by pestilential odours. For this abominable state of things, my uncle substituted straightaway and totally, the regime which is today standard in all large towns. (Philip 1995).

In developed countries, we are mostly unaware of the hydrology or of the hydrologists and engineers who have enabled us to have fresh, pure, potable water supplied from taps within our houses even in the most arid of locations.

While flooding persists in many areas, the constructions of levees and dams has mitigated the impacts of frequent flooding on human settlements. Dam construction has also contributed to flood protection and to continued supplies of water, particularly during droughts, and the diversion of water to arid areas has enabled the establishment of cities and irrigation areas that would not have been possible otherwise. The harvesting of groundwater also permits people to live and farm in arid areas and our reliance on groundwater, particularly deep groundwater, in such regions is increasing.

Irrigation is the major water user throughout the world. Without it, it would not be possible to supply the food and fibre necessary for the world's growing population. The application of hydrological science to irrigation has introduced new techniques, such as drip irrigation. Water, in drip irrigation, is fed through a small opening in a supply line directly to only the root zone of the irrigated plant at a suitable rate for the plant's transpiration demand. The whole soil surface is not wet, nor are there large excesses of soil water to drain and recharge groundwater. This has dramatically improved the efficiency of irrigation.

5. Controversies

Handling natural variability and the problem of scale

The inherent variation in space and time of the hydrological processes, particularly rainfall and streamflow, often defy exact prediction and measurement over many practical scales (Philip 1975). In addition, the use of point measurements to represent large-scale processes adds to the perceived variability. This variability is responsible for the extensive use of statistics throughout hydrology. The best we may be able to do is to admit that quite different laws may be appropriate at different scales so that prediction depends on the scale and the type of problem under consideration (Dooge 1997). For this reason, systems analysis, in which the hydrological system under study (say a river catchment) is treated as a black box with variable inputs and outputs, has been used, particularly for the study of larger-scale processes. The relationship between time-series of input signals (rainfall, temperature) and outputs (streamflow) for a subset of the measurement record may then provide a model or relationship between inputs and outputs. Given the inputs, this model can be used to predict the output over some other part or the whole of the measurement record (Young and Whitehead 1975).

A criticism of the systems approach is that, because it does not include a description of fundamental processes, it may be powerless to cope with temporal or spatial change (such as clearing within a catchment) unless system parameters have some physical meaning. A further criticism is that systems models, like statistical models, are difficult to falsify (Philip 1975). This has led to the notion that some problems in hydrology are 'trans-scientific', that is there are questions that can be asked of science that are unanswerable by science, because they transcend science (Philip 1975). Not all problems in hydrology are amenable to the scientific method. It has been suggested that attempts to model the impacts of global carbon dioxide increase on rainfall in particular locations using large process models may be trans-scientific. This is because the complexity of the number of coupled smaller-scale process models and the plethora of adjustable parameters they contain, defy attempts dispassionately to test large models.

Part of the problem with scale and variability arises from the use of point measurements, such as rainfall (in 200 mm rain gauges), to represent much broader scale processes. The problem of determining the relation between point measurements and broader scale hydrological processes is known as the scaling-up problem (see chapter 10). At least part of the trans-science issue is concerned with this problem. There are however some measures that are integrative, such as stream or groundwater chemistry, and provide information on broad scale processes, integrated over time and space.

Narrow focus

A criticism levelled at hydrology, and particularly engineering hydrology, is that it is narrowly focused, mainly concentrating on the protection and supply of water to human settlements. In this focus it has been spectacularly successful, but sometimes at the expense of the environment. Major water developments such as large dams, conceived as poverty alleviation projects, have in many areas served to increase the gap between the wealthy and poor and have caused the translocation of the poor to already crowded cities (Kaosa-ard 2003).

The case study of saline intrusion floodgates on the Mekong presents a particular example of problems that were unforeseen because of a focus on only one aspect of a regional problem, and a failure to appreciate other dimensions. In the wet season, the Vietnam's Mekong Delta is extensively inundated by floodwaters. This permits large-scale cultivation of rice, with sometimes up to three rice crops per year in northern areas. In the dry season, the diminished freshwater discharge of the Mekong allows saline seawater to intrude considerable distances into the Delta, particularly in the south western Ca Mau Peninsula. This limits the area available for dry season rice cropping. Because rice production was a priority of the Vietnamese government, studies of seawater intrusion led to a proposal to install major saline intrusion floodgates on main canals along the Ca Mau Peninsula and South China Coast (see figure 9.3). This scheme, called the Quan Lo Phung Hiep Project, was conceived to lengthen the growing season for rice from one to two crops in part of the saline intrusion affected area (figure 9.3) by decreasing saline ingress and increasing the flow of freshwater from the Bassac River during the dry season.

Construction of a series of 12 massive sluices or tidal floodgates in the Cau Mau Peninsula began in 1992 and was completed in 2001. Floodgates were installed on the major rivers and canals connected to the East and West Seas (figure 9.3). The project cost over US$12 billion and included the dredging of over 250 km of secondary canals. The sluice gates are between 5 and 25 m wide; they open automatically on the ebb tide and close on the spring tide.

At least three factors were overlooked in the design of this project. Brackish water canals are more productive in fish production than freshwater canals. Fish caught in the canals were the major protein source for subsistence farmers. One-crop rice farmers in the seawater intrusion area farmed shrimp in the dry season, a much more valuable crop than rice (White 2002).

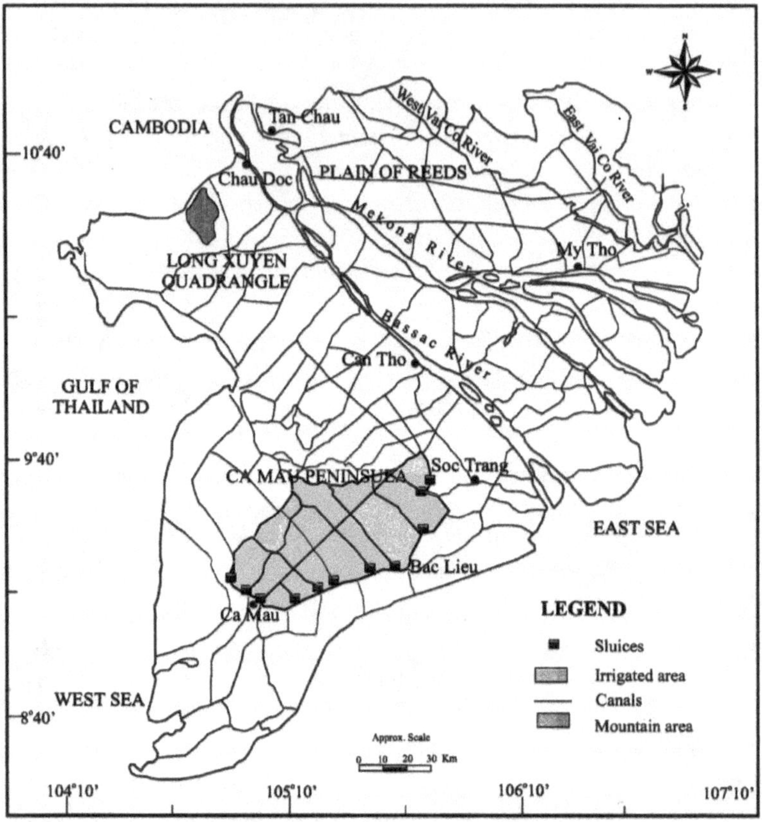

Figure 9.3. Location of the 12 saline intrusion sluices in the Quan Lo Phung Hiep Project in the Ca Mau Peninsula. The irrigation area protected behind the floodgates is shaded in grey (Ghassemi and Brennan 2000).

Some farmers in the western part of the peninsula had to abandon shrimp farming, which meant sharp declines in household incomes. Rice farmers in the eastern part, however, wanted to retain the salinity protection scheme because it had increased rice production and profitability. Salinity prevention measures also caused a decline in abundance of natural fisheries in rivers and canals. This resulted in declines in income earnings from capture fisheries, which were not only important income sources for poor households but also important protein sources. Significant ecological consequences of preventing tidal ingress into the study area have occurred. As more sluices went into operation, the rapid change in hydrological conditions had profound economic and social impacts on farmers' livelihoods. As a result, major conflicts over the operations of sluices have arisen.

Engineering education itself has also come in for criticism in reinforcing a narrow approach to complex problems:

> The message I took away with me was: 'All things are understood, and all a young engineer needs to know is what handbook to use.' The suppression of curiosity and removal of intellectual challenge made the course utterly boring, and its product brain dead (Philip 1998).

It is clear from the earliest records of irrigation on the Tigris and Euphrates rivers that irrigation in arid regions fails because of salinisation unless there are controls over recharge to groundwater (Ghassemi et al. 1995). It is therefore surprising that in the 21st century irrigation schemes without adequate drainage controls are still being constructed. This failure to consider the historical context for action (see chapter 2) is another continuing problem for narrowly focused hydrologists.

Hydrologists must broaden their outlook to include environments and societies. They have already made major contributions to water conservation schemes and to ways of recycling what was long regarded as waste water. They have improved the design of dam operating procedures so that there are options for better environmental flows from storages. In the re-engineering of the Florida wetlands, the Corps of Engineers of the United States Army are working to restore freshwater flows and distribution for greater biodiversity over the largest possible area.

Groundwater: out of sight, out of mind

Because it is out of sight, groundwater is mysterious to many people. Some consider that it exists as underground lakes or flowing rivers and that divining for groundwater with forked sticks and bent wire is widespread. Few understand how groundwater is recharged or the general intimate connection between groundwater and surfacewater. This lack of understanding is pervasive. The 1995 cap on surface water diversions from rivers in the Murray-Darling Basin in eastern Australia, designed to guarantee at least some environmental flows, did not include a limit on extraction from groundwater systems connected to the rivers. In many arid parts of the world groundwater is being mined — that is, it is being pumped out at a rate greater than its recharge rate. This causes groundwater levels to fall, dries up connected wetlands, decreases base flows in rivers and causes major seawater intrusion. Hydrologists have been unable to convince decision makers that this practice is, in the long run, unsustainable. In general, our understanding of groundwater systems, particularly the recharge rates of deep aquifers, is inadequate.

6. Unresolved issues

Global-scale hydrology remains an area of intellectual challenge both in terms of theoretical prediction and in measurement at appropriate scales. The ability to predict with reasonable certainty the consequences of global change remains a pressing issue. So too is the use of proxy methods to infer past climates and hydrological conditions well beyond historical measurements (palaeohydrology) to identify the full spectrum of impacts of land use change. There is an increasing challenge to persuade governments to continue the monitoring and analysis of water and associated land resources. At the other end of the scale, the interaction of microbiology and chemistry with the properties of water continues to provide challenges as well as solutions to water quality problems. Indeed it is at the intersection with other disciplines that exciting new challenges are emerging.

7. Boundaries of hydrology

Hydrology is limited by the assumptions of its underlying scientific and engineering disciplines, by its narrow focus on solvable problems and by incomplete knowledge, particularly in terms of handling scale and variability. Some water resource issues cannot be resolved by traditional hydrology because they involve questions beyond the realms of science and engineering.

Hydrology and water resource management

Hydrology is but one of many areas of knowledge contributing to water resource policy and management (see chapter 6). The broader and often more weighty contributors are identified by Bredehoeft (1997). In his hierarchy of factors contributing to decisions about water projects, legal aspects come first in terms of priority followed by political, economic (see also chapter 4) and, at the lowest level, technical considerations. This ranking is partly because of the incomplete nature of hydrological knowledge. It is little wonder that US hydrologist Myron Fiering should have observed 'As a scientist, I would rather be right; but as an engineer concerned with water I would rather be president' (quoted in Dooge 1997).

Perhaps the most hydrology can do is to 'inject some intellectual discipline into the republic of trans-science' (Philip 1975) in order to help societies make informed choices over the use, management and allocation of their water resources.

8. Further reading

Davies' (2003) *Fundamentals of hydrology* provides a non-mathematical entry into hydrology. For those who wish to delve further, Maidment's introductory chapter to his *Handbook of hydrology* (1992) and Dingman's *Physical hydrology* (2002) provide more mathematical details of aspects of hydrology. Raats et al. (2002) give an overview of broader issues in the application of mathematics to the biosphere. Buras (1997) presents some interesting views on the achievements of hydrology and challenges facing it. Eagleson (2002) describes the interaction of plants and hydrology. Hem (1992) provides an introduction to water quality, while Stumm and Morgan (1996) give a more comprehensive treatment. The causes and consequences of stream and land salinity are discussed in detail in Ghassemi et al. (1995). Finally Philip (1995) is a highly entertaining vignette of one of the forgotten heroes of hydrology.

10 Mathematical and Statistical Sciences

M.F. Hutchinson

1. The origins of mathematics

Over the centuries, mathematics has developed a mystique that has set it apart from its companion sciences. It has been called the queen and the servant of science.[18] It can be viewed in abstract logical terms as a vast and intricate array of abstract tautologies, having immense inner beauty to its dedicated practitioners, but without any necessary practical import. Yet it has been found to be an enormously powerful tool for describing the natural world, eliciting subtle connections and implications both within and between natural systems. Some say that 'mathematics has an objective existence, providing structure to this universe, independent of and transcending the existence of human beings, or any beings at all' (Lakoff and Núñez 2000).

The success of Einstein's special and general theories of relativity encouraged the view of mathematics as an independent source of truth. When Einstein's theories were found to agree with observations they revolutionised popular and scientific thought in the early 20th century. Yet they were built entirely within an existing mathematical paradigm. Substantial technical additions to existing mathematical theory were made in the process but no fundamental tenets of mathematics were cast aside.[19] Einstein's theories represent one example of what appears to be a natural goal of civilisations to understand and codify their observations of the world.

From the earliest times mathematics has been motivated by intensely practical concerns. The agricultural economies of Egypt and

Mesopotamia needed a reliable calendar to overcome the vagaries of mythology and haphazard observation. The Sumerians are known to have dated the beginning of their year from the vernal equinox by 5700 BC and by 2500 BC Sumerian merchants were familiar with weights and measures supported by an arithmetic based on a sexagesimal (base 60) number system. Early measurement of land areas and engineering by the Greeks and Egyptians similarly required a number system and basic algebra. Moreover, arithmetic in support of astronomical calculations appears to have been developed independently by several cultures, including the Chinese, the Mayans of Central America and perhaps the Minoans of Crete.

Mathematics is a fundamental discipline that is essential to human experience. It has no independent truth separate from human thought. It arises in part from basic arithmetic structures in the human brain and is based on an elaborate array of conceptual metaphors constructed by the human mind (Lakoff and Núñez 2000). These metaphors are the means by which the abstract is made concrete. They can act as a 'resting place' to sum up our understanding of particular concepts and allow the mind the freedom to build on these concepts to further explain the human experience. It could be argued that the success of mathematics in codifying the natural world is due to the fact that it has been constructed by humans to do just that. Concepts that succeed in enhancing our understanding are retained while concepts that do not are forgotten.

2. The role of mathematics and statistics in environmental modelling

The conceptual analysis above (which may not be shared by all mathematicians) provides an empirical context for the role of mathematics in environmental modelling. Until recent times, the emphasis has been largely a reductionist one of constructing and testing models of key environmental processes. Models are amended or rejected when their outputs do not accord with observations. The best models can add to our understanding of these processes. They also enable effective simulation of environmental processes on a computer, often as part of a larger interacting system. Using these models and systems, together with their supporting data, specific courses of action can be identified to reduce environmental hazards and to enhance the welfare of the living world.

Effective model construction depends as much on the art of the modeller and the nature of the environmental process under investigation as it does on the availability of mathematical concepts and tools. Mathematical sophistication is not a goal in its own right and a simpler model is often preferable, especially if it matches observed behaviour as well as a more complex model. It is even more preferable if the simpler model is in fact closer to the modeller's perception of the environmental process. Statistics also has a key role in helping the modeller to assess the appropriateness of a proposed mathematical model by analysing and assessing both model error and data measurement error.

Mathematical statistics

Mathematical statistics is generally seen as an amalgamation of the disciplines of probability theory and statistics (Larsen and Marx 1986). Probability theory arose in the mid-1600s from observations of odds of success in games of chance by Pascal, Fermat and Huygens. It was firmly established with the work of Jacob Bernoulli in the early 1700s that included a theorem on the probability of repetitions in a simple chance experiment. Statistics also arose in the mid-1600s from interpretations by John Graunt of collections of mortality data. Succeeding work by Edmond Halley in the 1700s on mortality tables and life expectancy established statistical inference as a viable and practically useful science.

The two disciplines were unified by Pierre-Simon Laplace in 1812 in his *Théorie Analytique de la Probabilité*. This book discussed the application of probability to the theory of observational errors, in particular identifying the utility of the *normal distribution*. This in turn established that the theory of probability could be profitably applied to empirical data in the physical sciences. It became clear that the observed behaviour of some physical systems could be cogently summarised, and to some extent understood, in mathematical statistical terms by making an appropriate assumption of randomness. Key ensuing developments were regression modelling developed by

Francis Galton in the late 1800s and the associated theory of statistical inference developed by R.A. Fisher in the early 1900s.

It is now acknowledged that mathematical statistics has a wide range of applications in agriculture, biology, demography, economics, geography, management of natural resources, medicine, meteorology, politics, psychology and sociology. These fields cover the whole spectrum of environmental modelling. Statistics not only plays a fundamental role in calibrating and validating environmental models. It can also generate models of variability in its own right. These models can provide insight into the structure and complexity of environmental processes at various scales and predict aspects of these processes over time and space.

Key concepts for environmental models

A single chapter cannot pretend to provide a comprehensive treatment of the vast range of mathematical models. Thus this chapter uses examples to discuss some key concepts for mathematical environmental modelling. Temporal and spatial dimensions are fundamentally important for environmental modelling. Accordingly, both temporal and spatial models are described in their historical context.

Temporal models are typically applied to processes at a single point in space. They can also be applied to large spatial entities, such as whole river catchments, or even the whole earth, if spatial variation within the entity is ignored. In this case properties of the entity are aggregated, typically by taking spatial averages, so that the spatial entity is acting as a single point. Such models are called *lumped parameter models*. They can provide useful descriptions of the dynamics of large systems as long as their limitations in aggregating spatial variability are understood.

What sorts of models?

Sections 3, 4 and 5 below provide examples of three different types of models used in environmental studies. Section 3 explores a model for continuous population growth over time generated by differential calculus. The basic process of model construction and evaluation is described, including the role of statistics in this process. Section 4 considers a model of precipitation at a single point. This model is also in the time domain but has a very different form. It is a simple *marked point process*. It shows how a simple statistical model can explain observed temporal behaviour for an environmental process. Section 5 discusses spatial multivariate models fitted to spatially distributed environmental data. These models have a great range of applications because of the inherent spatial nature of many environmental processes. Spatial models are used to describe landscapes and the processes that they support. Just as temporal models can ignore spatial variability, spatial models can ignore temporal variability. The actual terrain surface underlying a landscape is one that can be relatively static over shorter timescales, although not over geological timescales, and can be described by a spatial model that is fixed in time.

Ultimately all environmental processes vary in space and time. Daily weather is a classical example. The construction of models that effectively address simultaneous variation in space and time remains a challenging field. The natural limits on complexity of spatial models apply all the more to spatio-temporal models. The chapter concludes with a discussion of complex systems that, by directly incorporating *non-linear interacting systems,* have the potential to generate a fundamental shift in environmental modelling.

3. An environmental growth model

This section discusses a simple population growth model based on differential calculus, a tool developed independently by Isaac Newton and Gottfried Leibniz in the late 1600s. Differential calculus made it possible to represent arithmetically rates of change with respect to time and space. It was of fundamental importance in the development of modern physics and mathematics. It permitted Newton to succinctly account for Kepler's observations of the motions of the planets with his inverse square law of gravitation. He similarly accounted for Galileo's observations of the motions of falling and rolling bodies with his three fundamental laws of motion. Differential equations also have applications in all sorts of domains including physics, biology, economics, medicine and many others.

Model formulation

A classical application of differential calculus to environmental modelling is for calibrating rates of *growth* and *decay*. For a natural population with no predators and no limits on resources it is reasonable to expect the rate of population growth to be approximately proportional to the size of the population. This can be written:

$$\frac{dP}{dt} = kP$$

where the left hand side denotes the rate of growth of the population, that is, the *derivative* with respect to time t of the population P, and the right hand side expresses this rate of growth as a proportion k of the population P.

Underlying this equation is the notion that the population P is a *function* of time t. Since time is usually thought to be *continuous* there is a tacit assumption here that the population is also a continuous function of time. This is not strictly true for a population consisting of *discrete* individuals but it is a good working approximation for a large population. In fact this approximation is often made in physical modelling. The reverse approximation is also made when continuous equations are solved on a digital computer where continuous real numbers are represented as discrete quantities.

Solution and model refinement

The fundamental role of the *integral* calculus is to solve differential systems. In the case above, if the proportion k is assumed to be constant over time, in reality a dubious assumption, the equation can be integrated to give the equation of Malthusian exponential growth given by

$$P = Ae^{kt}$$

where e is a fixed constant and A is the size of the population at time $t = 0$. This equation is defined for all values of t. The population becomes infinitesimally small in the past for large negative values of t and exponentially large in the future for large positive values of t.

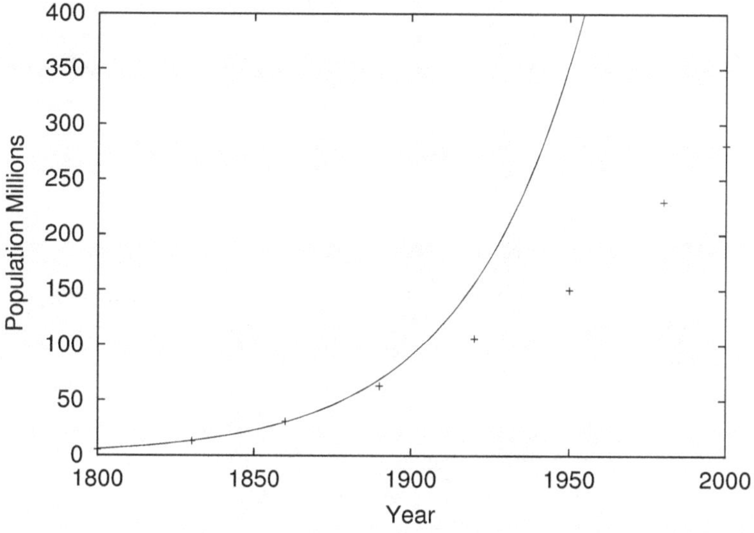

Figure 10.1. Exponential growth model of the population of the USA.

The fallacy of assuming a constant growth rate for the population of the USA from 1800 to 1980 is shown in figure 10.1. Here the exponential growth equation can be seen to model the population reasonably well over the 1800s but clearly breaks down in the 1900s. The two World Wars, the Depression, varying immigration rates and ultimately, pressure on natural resources, all play their part in violating the assumption of a constant growth rate. Measurement error also plays a role in the departure of the fitted model from the data. Earlier population measures are likely to be less accurate because of various logistical difficulties.

The clear inadequacy of this model suggests that the simple differential equation should be modified to take account of at least some of the listed constraints. A common approach with growth models is to assume that there is a maximum sustainable population M and write

$$\frac{dP}{dt} = kP(1 - P/M)$$

This differential equation is known as Verhulst's logistic law of population growth (Finizio and Ladas 1982). In this case P grows essentially exponentially at rate k while P is small compared to the posed maximum population M. However, the growth rate eventually declines to zero as P approaches the maximum population M.

Calibrating the model

The integral of the logistic equation gives P as a function of the growth rate k, the maximum population M and the population A at time zero. This function is *fitted* to the same data shown in figure 10.1 and plotted in figure 10.2. The process of fitting the equation to the observed data is termed model calibration. It involves finding the unique values of the unknown *parameters* k, M and A that, in this case, minimise the differences between the fitted function and the observed data. Statistical considerations immediately arise in this process. These include making an assumption on the nature of these differences. The differences are measured in the same terms as the population is measured. Thus they are the vertical differences between the points and the fitted function in figure 10.2. This framework can be summarised by writing an equation for each observed population value P_i at time t_i. in terms of the posed logistic model. This may be written

$$P_i = f(t_i, k, A, M) + \varepsilon_i$$

where the function f is the integral of logistic differential equation above and ε_i is the difference between the logistic model and the observed population value P_i at time t_i. This equation exhibits the fundamental paradigm of *regression* modelling where the observed data are split into a well defined function, also called the *signal,* and the essentially random differences ε_i, the *noise*. The noise can arise from two fundamental sources – deficiencies in the model and measurement error. Both of these can make a significant contribution to the differences between the fitted model and the data.

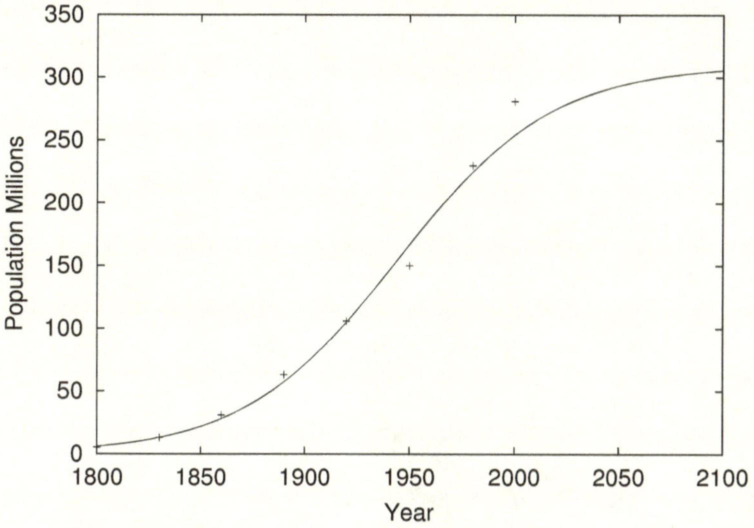

Figure 10.2. Logistic growth model of the population of the USA.

The unknown parameters can be estimated by maximising the *likelihood* of the fitted function, a concept formalised by Fisher in the early 1900s (Kendall and Stuart 1979). The likelihood depends on an assumed statistical distribution for the differences. A standard assumption is to assume that the differences (also called *errors*) are distributed according to the *normal distribution* of Laplace. This statistical distribution commonly arises via the *central limit theorem* when dealing with quantities obtained from some sort of accumulative process (Rohatgi 1976, Larsen and Marx 1986). If the errors are assumed to be distributed according to the normal distribution then the likelihood is maximised by minimising the sum of the squares of the differences of the model from the data.

However, many environmental processes, and their associated errors, are naturally non-normal. Thus precipitation and streamflow measurements are highly *skewed*, with many small or zero values and few large values. This commonly arises with physical quantities that are naturally non-negative. This skewness can be removed by a suitable transformation of the data, achieved in the case of precipitation by raising the precipitation values to a small power or in other cases by taking logarithms.[20] Alternatively the form of the distribution of the errors may be known and the likelihood can be calculated accordingly.

If suitable action is not taken with highly skewed data then the accuracy of the fitted model can be compromised by fitting large data values too closely and not fitting small values closely enough. Significant data errors can also be overlooked. For the population model here the population values are naturally positive and approximately exponential in character. The logarithm of the model was therefore fitted to logarithms of the measured population values. The values of the parameters k and M were given by k=2.73±0.03% and M=310±30million.

Model testing and interpretation

Model testing is an essential aspect of environmental modelling. The fitted model can be tested by using the model to make predictions about information that has not been used in the model calibration process. Agreement with the prediction of a model can only confirm consistency of observations with the model, never assert its ultimate truth. However, continued agreements with model predictions will lead to wider acceptance and use of a model. Serious disagreement of data with a model prediction generally leads to rejection or refinement of the model or to correction or deletion of erroneous calibration data. It is also important to check the veracity of the test data.

Model testing is most commonly achieved by withholding data from the model calibration process and then checking model values at the withheld data points. For example, in the model of the US the population the actual population in the year 2000 of 281 million was not used in the model calibration. The value of the fitted model for the year 2000 is 254±20 million. This underestimates the true value and suggests, not surprisingly, that the logistic model is by no means a complete model of the population growth of the USA, despite its improvement over the exponential model. The suggested maximum population is rather low in the light of recent census data that suggest that the maximum USA population could be around 380 million.

Extensions and applications

One extension to the above formulation is to make the growth parameter k of the logistic model a *random variable* instead of a constant. This would appear to be a realistic assumption. It places the model in the field of *stochastic differential equations*, a major area of study in its own right (Kloeden and Platen 1995).

A second extension is to apply the logistic model to plant growth. In this case the growth parameter k can be reasonably assumed to vary systematically over the year according to climatic and soil moisture conditions. Such systems cannot be solved explicitly. However they can be addressed by numerical simulation using the relevant climatic and soil moisture inputs and *numerical integration* to calculate plant biomass as a function of time.

A third extension is to use interacting differential systems. Hydrological systems and many biological systems can be formulated as differential systems that include feedback loops (Finizio and Ladas 1982). Non-linear interacting systems can give rise to complex behaviour that is not usually amenable to complete analytical solution.

4. A physically based temporal model: precipitation

Precipitation is a key driver of natural and human systems. It supplies moisture for vegetation and agriculture and supplies drinking water for most animal species. Rivers are also fundamental drivers of landform erosion and deposition. The highly variable nature of accumulated precipitation at any place has posed a particular challenge for model developers. It has a fundamentally different character to the continuous population growth described above. It is normally observed at regular intervals of hours, days, weeks or months. At any of these timescales the observations exhibit discontinuous behaviour, with wet periods associated with specific rainfall events interspersed with dry (zero precipitation) periods. The timescale of observation affects the nature of models that can be posed and the sorts of dependent processes that such models can help assess. Relatively fine timescales of hours and days are required for processes such as soil erosion and floods. On the other hand, the broader monthly timescale permits use of relatively simple models and can be used in assessments of broad scale natural processes such as accumulated plant growth and drought (Hutchinson 1995).

The *compound Poisson model* (Ross 1983) is a statistical model that aims to mimic the event-based nature of precipitation. It assumes that rainfall events occur as discrete random amounts at instants in time, which are also separated randomly in time. The time separating each event is assumed to be distributed according to an *exponential* distribution, giving rise to a standard Poisson occurrence process (Feller 1967). This distribution has a single parameter that can be specified by the mean number m of events per month. Similarly, the amount of precipitation is assumed to have an exponential distribution that can be parameterised by the mean rainfall per event. The overall model therefore has just two parameters. The modelled instantaneous duration of each event is unrealistic at daily and finer timescales, but event duration has little consequence at the monthly timescale.

Despite its simplicity, this model can explain one of the key features of observed rainfall behaviour. A standard measure of rainfall variability is the *coefficient of variation,* the ratio of the standard deviation to the mean. It is commonly observed that the coefficient of variation of observed precipitation totals is large in arid areas. Thus arid Australia is known to have a very variable rainfall regime. The compound Poisson distribution offers a simple explanation of this phenomenon. The coefficient of variation CV of monthly totals for the compound Poisson distribution is given by Figure 10.2.

$$CV = \left(\frac{2}{m}\right)^{1/2}$$

where m is the mean number of events per month. It thus depends only on the rate of occurrence of rainfall events, not their amounts and CV increases as the rate of occurrence gets small. Observed monthly rainfall coefficients of variation are known to have this form. Thus the high variability of precipitation in arid areas is shown to be directly related to the low rate of occurrence of events.

5. Spatial multivariate models

Environmental quantities are naturally distributed over space so accounting for the spatial distribution of environmental variables plays a key role in environmental modelling. Biophysical variables such as climate, soil and terrain are the primary spatially distributed determinants of plant growth. Dependent quantities such as natural vegetation, agricultural productivity, soil erosion and human and animal populations all have spatial dimensions that can be addressed by making explicit links to controlling spatial processes. The type of spatial analysis that is discussed here is usually termed *geostatistical*. It is applied to data that have been sampled at points across the landscape.

The broad aim of geostatistical analysis is to identify the nature of the spatial coherence of such data and use it to estimate (or interpolate) the complete spatial distribution from the incomplete sampled points. A significant body of related work on multivariate smoothing spline methods also arose in the late 1970s, largely championed by Wahba (1990).[21] The methods are formally equivalent, and have similar accuracy, but tend to differ in practice (Hutchinson and Gessler 1994).

A *spline* was originally a long, narrow, and relatively thin piece or strip of wood or metal (for example, a slat) that curves when weight is applied to it. Mathematics has borrowed this word to apply to a flexible approximation of an unknown relationship fitted to a data set. A *spline function* is used to approximate mathematically a function or an unknown relation between two or more variables (for example, precipitation as a function of horizontal position and elevation). Since precipitation has a strong dependence on elevation, such spline functions can be well determined from widely scattered data points. The flexible function works like the original slat to reflect weightings. The idea is to posit theoretically enough interpolating points to 'smooth' the known data, and make for a function that can be manipulated mathematically. Such functions can then be used to generate images (in two or higher dimensions) of the relations between environmental phenomena.

Limit on complexity of multivariate spatial models

There is a fundamental limit on the complexity of multivariate spatial models fitted to data. The process of adding additional predictors, such as elevation, to a multivariate spatial model cannot be continued without limit. Additional predictors rapidly enlarge the space in which the model is fitted. This is called the *curse of dimensionality*. Multivariate spline functions can have quite definite constraints imposed on their structure[22] (Wahba 1990). These can be effective, provided the constraints on structure are consistent with the process being modelled.

6. Spatio-temporal models

The subject of spatio-temporal models is challenging in both theoretical and practical terms. Model structure needs to reflect separate temporal and spatial behaviour, as well as interactions between space and time. The chief practical limitation relates to obtaining enough data to fit a complex model. Efficient management of large spatio-temporal data sets can also impose practical difficulties.

Three approaches may be identified. For phenomena that have little of no interaction between spatial and temporal behaviour, the spatio-temporal model need be no more than a succession of spatial models over time. The spatial models at each time step can then be calculated independently from the spatial models fitted at other time steps. These can be appropriate for stochastic weather modelling if the time steps are relatively large, such as months or years.

At shorter time steps, transport processes tend to give rise to significant interactions between spatial and temporal behaviour. For phenomena that are continuous in space and time the multivariate normal distribution (Rohatgi 1976, Davis 1986) is a common choice. A particular advantage of this approach is that the interactions between spatial and temporal variables are completely specified by the corresponding correlations. Such an approach has been used to develop a model of atmospheric turbulence.[23]

For event-based processes in space and time, subsidiary structures can be attached to the basic spatio-temporal occurrence process to match observed structure of precipitation events. Such models include *Poisson cluster process models*. While these model structures can be appealing, these models can be difficult to calibrate because there may be insufficient data to accurately calibrate the parameters associated with the subsidiary event structures.

7. Complexity theory

Traditional mathematical models are often successful in describing sub-systems in quite precise terms, but are less successful in describing higher levels of organisation. Cohen and Stewart (1995) argue that this complexity arises from relatively simple interactions of entities 'one level down'.

The predator–prey model, developed independently by Lotka and Volterra in the 1920s, is a much-quoted example where complexity can arise from the interaction of very simple systems. Suppose we have a population of predators *P* (foxes) and a population of prey *H* (hares). The equations may be written

$$\frac{dH}{dt} = r_H H - aHP$$
$$\frac{dP}{dt} = -r_P P + bHP$$

where r_H is the natural rate of increase in the prey and r_P is the natural mortality rate of the predators. The term aHP represents the decrease in prey due to predation, proportional to the product HP of prey and predators. The term bHP represents the growth in the predators due to predation, also proportional to the product of prey and predators. According to these equations, if there were no predators then the prey population H would simply increase according to the Malthusian exponential growth law described earlier. Conversely, if there were no prey then the predator population P would undergo exponential decline.

The terms accounting for the interaction of predators and prey dramatically change this behaviour. It can be shown that if the interaction terms are both non-zero then the two populations will oscillate indefinitely (Finizio and Ladas 1982). Volterra invented this model to explain the increase in the observed proportion of selachians (sharks, skates, rays) in fishermen's catches in the northern Adriatic Sea during World War I. It was thought that the change was due to the reduction in fishing during the war. Since fishing reduces the number of selachians and fish, fishing reduces the rate of growth of the prey (the fish). It also directly increases the mortality of the predators (the selachians). The equations predict, surprisingly, that increased fishing rates can lead to an increase in the *average* numbers of prey while the average numbers of predators decrease, as expected. This is precisely what was observed in the fishermen's catches. Although this model is an oversimplification of a real predator–prey system, it does explain subtle observed behaviour – a key aim of reductionist mathematical modelling. Such models can also reveal features, such as mean populations, that are not directly observable. These features can be of direct use in applications, such as assessing expected incomes of fishermen.

Simplicity from complexity

Real predator–prey systems are more complex but they do exhibit stable cycles that the system will tend to return to over time. Such stable cycles are called *attractors* in dynamic systems theory. Other dynamic systems, not necessarily any more complex in formulation than the predator prey model, can exhibit *chaotic* behaviour because of high sensitivity to initial conditions. This is the *butterfly* effect observed by Massachusetts Institute of Technology meteorologist Edward O. Lorenz when investigating numerical weather prediction in the 1960s. Even for chaotic systems attractors still exist but have *fractal* structure, meaning that they have structure at all scales. Attractors are an *emergent phenomenon* in dynamic systems that could well have a role in underpinning statistical characterisations of complex phenomena (Cohen and Stewart 1995).

Multiple attractors can give rise to rapid change in systems subject to incremental change in forcing conditions. This is the phenomenon recognised as *catastrophe theory* in the early 1970s. It is thought, for example, that the global climate system has two relatively stable attractors, the mild climate currently being experienced and the climate of the ice ages. Leaving aside the major ice age cycles that may be related to perturbations in planetary motion, other rapid large-scale changes in the global climate system that have occurred in the past are unpredictable in reductionist terms, for two essential reasons. One is that the global climate system is so complex that it is impossible to model all of the component interacting systems. Secondly, even if these systems were known, it is impossible to know their precise state at any one time. Given the sensitivity to initial conditions, predictions of the state of the global climate beyond a relatively short lead time are inevitably bound to fail.

The challenge of complexity theory is to find simplicity in such complex dynamic systems that can be put to good purpose. Progress to date has been more in conceptual understanding than in providing quantitative solutions. The approach differs fundamentally from reductionist modelling by dealing with interacting systems as a whole rather than attempting to assemble them from their component parts. Further quantitative progress in complexity theory would not supplant existing reductionist modelling, but would complement it in ways that have yet to be discovered.

8. Further reading

Lakoff and Núñez (2000) provide a modern perspective on mathematics and its foundations. Larsen and Marx (1986) provide a thorough account of mathematical statistics and its applications in historical context. Feller (1967) is an older classic that communicates much of the excitement of statistics and probability theory. Davis (1986) describes statistical modelling in the earth sciences, considering both temporal and spatial models. Chilès and Delfiner (1999) provide a thorough account of geostatistical modelling, while Schimek (2000) explains spline-based approaches to multivariate modelling. Maidment (1993) provides a comprehensive survey of hydrological models and Von Storch and Zwiers (2002) survey statistical atmospheric models. Howser et al. (2004) provide an account of how spatio-temporal statistical models can be applied to the assessment of the impacts of projected climate change.

This chapter has not been able to deal with issues of scale that arise when functions are decomposed into hierarchical components, such as those afforded by Fourier series. Fourier series decompositions have much use in simplifying problems for both theory and applications. For an introduction, see a modern engineering text such as Kreyszig (1999). Stewart (1989) provides an excellent general account of chaos theory in the context of modern science.

11 GEOGRAPHY

Richard Baker

1. Introduction

Geography's central concerns are the complexities and uncertainties of human–environment interactions. While a subject like physics can have formulae that can accurately predict the behaviour of planetary orbits, geography's insights are usually too complicated and too context-specific to be so eloquently summarised. Individual human behaviour and the nature of human interactions are so much more unpredictable than the path of our planet through the solar system. Likewise, cultures mediate individual interaction with the environment in ways that are impossible to reduce to universal equations. Moreover, the feedback loops between people and the environment are extremely complex and impossible to predict accurately.

While human–environment interactions cannot be reduced to formulae it is crucially important that we develop a better understanding of them. Geography has played a crucial role for centuries in assisting cultures around the world to understand their relationship with their environments. This long tradition of exploring human–environment interactions is of great contemporary relevance given the current dimensions, complexities and potential catastrophic outcomes for our planet of these interactions.

2. What is geography?

Like history, geography is something everyone does. Just as we all try to make sense of our personal and family past, we all use maps of one form or another in our everyday lives. Most of us have also studied geography in schools. Unfortunately the geography that many of us did at school was limited to rote learning an array of geographic facts. Such 'butterfly collecting' approaches include learning what is the capital city of distant countries, the height of the world's highest mountain peaks or the length of the longest rivers. Only some of us were fortunate to have been taught geography in a more integrative, in-depth and meaningful way to reveal the discipline's enormous potential for helping us understand the world around us. Such geography, rather than ignoring the complexity of human–environment interactions, has learning about these complex issues as its central focus.

Not surprisingly the general lay view of geography reflects the simplistic way it has been taught in many schools. A slightly more sophisticated popular view of geography is that it is about exotic people and places. The popularity of Discovery Channel TV and the *National Geographic* magazine reveal that 'exotic people and places' geography has enormous popular interest. The reach and influence of such programs and magazines are enormous and clearly play a crucial role in shaping geography's popular image (Knox and Marston 2001: 3).

While the general community has a strong (albeit simplistic) understanding of what geography is, professional geographers cannot agree on a single definition of geography. Such disagreement is not surprising given the discipline's breadth of interest, capacity to provide intellectual bridges to a wide range of disciplines, and the extraordinary diversity in methods, approaches and subject matter. As this chapter highlights, geography's breadth and links with other disciplines are strengths particularly pertinent to dealing with crucial human–environmental issues facing this planet. Significantly these strengths are the basis for 'adaptive management' – generally considered to be the best approach in dealing with variability and uncertainty in natural systems.

Geographers are interested in the intimately interconnected concepts of place, space and scale. Events, phenomena and actions all occur in space, and are all fruitfully examined spatially. Places need to be understood in terms of their locations in space and the scale of various phenomena is a key to understanding the interconnections between places. Human–environmental interactions have a spatial dimension, and geography's interest in location and the distribution of resources, people and activities is crucial in understanding these interactions. This focus has led some to define geography as being about 'who gets what, where and when'. Such a definition gives an inherently political dimension to geography as such questions of allocation raise numerous equity issues. For example who (or what processes) decide who gets what, where and when?

While geographers do not share a single common method, some aspects of their work are common to all. They study the complex interplay between people and the environment and how the two shape each other. Geographers have at times oversimplified this complex interplay. Geography has gone through both culturally deterministic phases that overly asserted the dominance of culture over nature, and environmentally deterministic phases that overly stressed the role of the environment determining human behaviour.

Both of these extremes miss the mark and fail to take advantage of geography's great powers to integrate human–environment issues. British and North American geography in the 19th and early 20th centuries went through a very environmentally deterministic period. 'Natural regions' defined by natural features of topography, climate and vegetation were seen to give rise to characteristics of human behaviour. It was argued, for example, that 'invigorating climates' such as those of Britain and North American were more conducive to clear thinking than tropical environments.

3. History of geography

Geography can be dated back to ancient Greece, Rome, China and Arabia. Significantly in each case, geography was part of these countries' respective eras of discovery, and geography played a crucial role in mapping the resources (both human and natural) of the 'new worlds' that these cultures encountered. Many regard Strabo (60 BC to about 21 AD) as the first true geographer. His 17 volume *Geography* was written between 9 BC and 18 AD. We would today describe his work as regional geography. His *Geography* maps out local relationships between nature and society. Ptolemy (90–168 AD) was another key early Greek geographer (he was also known as an astronomer). In his eight-volume *Geography* Ptolemy set himself the ambitious task of 'developing a comprehensive view of the world' (Knox and Marston 2001: 10).

Significantly, like its ancient predecessors, modern academic geography developed in an age of discovery. The British Royal Geographic Society was established in 1830 with a major focus on promoting exploration. This focus continues to the present. In November 2003 their journal *Geographical*, for example, had a special feature that outlined the 10 steps to successful expeditions. In the 19th century the Royal Geographic Society played an active and highly successful role in lobbying for the establishment of university geography departments. By the end of the 19th century university chairs in geography at many of the most prestigious universities and numerous geographical societies had been established (Stoddart 1986: chapter 3).

German and French geography also have had a major influence on modern geography. The French school of regional geography led by the work of Vidal de la Blache placed great emphasis on the personality ('genres de vie') of regions as a unifying theme that provides explanations of both regional character and variations. Perhaps the most influential German geographer was Ratzel. The first volume of his *Anthropogeographie* played a major role in the trend towards environmental deterministic by promoting the idea that the physical environment is a key force in shaping societies. The fact that his second volume charted the opposite process of societies shaping the environment was largely forgotten until the rise of Carl Sauer who had a major influence on cultural geography.[24]

Sauer's 1925 paper 'The morphology of landscape' was enormously influential in elevating landscape as a central focus of geography. In this paper he also pioneered cultural geography by developing the concept of cultural landscapes that are made by cultures over time, modifying natural landscapes with layer upon layer of cultural changes. Sauer also emphasised the virtues of geography's field-based methods. In Sauer's view geography is something that you do through the 'soles of your feet'.[25] Sauer's work has inspired many others to stress the importance of developing skills in reading the landscape, as one would read a book, for signs of human impacts and meanings. 'Land literacy' skills are practical as well as theoretical. In contemporary Australia, the landcare movement has sponsored community environmental monitoring programs. Landcare has encouraged rural landholders to share ideas with each other on how to read the warning signs of impending land degradation on their properties.[26]

Sauer's 1941 *Foreword to historical geography* was seminal within historical geography. In this, he presented a convincing case for the centrality of historical geography within geography by demonstrating that 'historical regional studies are in the best and oldest geographic tradition'. He noted, 'I should not be interested in historical geography or in human geography except as helping to understand the differentiation of cultures, and I cannot get understanding of this sort except by learning the ways and devices men have used for making a living out of their homelands'. He also argued for the broadest possible context for historical geography, calling for it to 'admit the whole span of man's existence … from all the earth in all the time of human existence' (Sauer 1941a: 11, 16, 24). Historical geography provides a powerful tool to chart the complex factors responsible for creating current environmental problems.

Elsewhere, Sauer stressed how important it is to consider the views of former inhabitants when trying to understand landscapes. In his paper 'The personality of Mexico', for example, he notes that 'the two most important things to know about Mexico are still the patterns of life that existed before the coming of the white men and the changes that were introduced during the first generation or two of the Spanish period' (Sauer 1941b: 354). Sauer's use of the term 'personality' in the title of his article reflects his intellectual borrowings from the French school of regional geography.

In more recent decades cultural geography has focused more explicitly on how different groups and societies perceive and use the environment. Cultural geography presents a perspective of particular value to 'settler nations' such as Australia and New Zealand where the recent arrival of non-Indigenous people radically changed a landscape. The interest of cultural geography in the power of ideas to shape landscapes helps us see the degree to which the 'cultural baggage' that people creating settler societies brought with them dramatically changed the landscapes. The wholesale importation of inappropriate land uses from the other side of the planet played a major role in the massive land and water degradation problems currently facing both Australia and New Zealand. The dominant European heritage of the settlers, and the maintenance of many European attitudes not relevant to their new homes, led them to see naturally recurring environmental events as aberrations to fight rather than live with. For example, while Indigenous Australians lived with droughts and bushfires, European settlers and their descendants tended to define these events as disasters to battle.

Historical geography as advocated by Sauer is an integrative subdiscipline that brings together the social and biophysical sciences. For much of the history of academic geography it has been split into human and physical components, and this has weakened the discipline. Such divisions have often been imposed externally by university administrators who have been unable to cope with a discipline that is both an art and a science. Geographers themselves have often perpetuated the division and failed to take advantage of their discipline's enormous integrative capacities. The rest of this section briefly outlines two other integrative subdisciplines — applied geography and regional geography.

Applied geography informs contemporary environmental issues. In a manifesto for an applied people's geography, David Harvey (1984: 7) notes that geographers 'cannot remain neutral'. In documenting human–environment relations geographers are always revealing issues of equity and inequality. This means geographers have a clear ethical choice — to ignore the implications of their results, or to take an 'applied geography' approach and try to do something to rectify inequitable situations. The choice is a political and moral one.

There is a long tradition of applied geographic research that engages with communities in explicitly interventionist ways. The links between such applied geography and planning are very strong. Indeed, one author argues that 'planning could more fruitfully be seen as a form of applied geography in all its diversity, complexity and contradictions' (Huxley 1999: 129).

Another significant area of applied geography has been the engagement of geographers in debates about development and conservation issues. Geographers with their interest in both social and biophysical processes have been well placed to point out the crucial, but often ignored, connections between the two sets of processes. Such work has also highlighted the need to look for deep underlying causes and not just symptoms, when developing policies to address environment problems. A good example of this approach is the work of the influential British geographer Piers Blaikie on why soil conservation policies usually fail. His work has enormous relevance to soil conservation approaches in both developed and developing countries. As he points out, it is essential to acknowledge that soil conservation is 'as much about social processes as physical ones, and the major constraints are not technical (in the agricultural engineering sense), but social' (Blaikie 1985: 50). Applied geography approaches (such as Blaikie's) that have managed to engage with policy-making have had many positive outcomes for communities around the world.

Regional geography, long out of fashion in academic geography, also has an important role to play in addressing contemporary environmental issues. Linked with applied geography it provides a powerful tool both in understanding complex local issues and in moving towards solutions. For example, such an approach is essential in addressing the acute water and land degradation facing many communities around the world. An applied regional geography approach can provide a sensitive and nuanced understanding of the interactions between a wide range of local issues such as groundwater, geomorphology, social and economic geography and historical geography.

4. Closely related disciplines

Geography has strong links with two of the other disciplines discussed in this book, history (see chapter 2) and anthropology (see chapter 3). (It also has strong intersections with environmental earth sciences, hydrology and policy.)

History

The traditional separation of history and geography has had ramifications for both disciplines. History has been poorer for its tendency to see the earth just as a stage for history without reflecting that the stage itself is alive. In turn, geography, especially human geography, has been limited by its general lack of historical depth (see chapter 2). As geographer Donald Meinig (1986: xv) notes, the connections between geography and history are potentially profound:

> Geography is not just a physical stage for the historical drama, nor just a set of facts about areas of the earth; it is a special way of looking at the world ... Geography and history are not only analogous, but complementary and interdependent, bound together by the very nature of things. This relationship is implied by such common terms as space and time, area and era, places and events — pairs that are fundamentally inseparable.

Anthropology

Modern anthropology and geography have a common heritage dating back to the age of discovery when they respectively documented new people and new lands as Europeans encountered them. As such both disciplines played a significant role in western colonisation (see also chapter 3). The two disciplines came closer together during phases of environmental determinism in the 1920s and 1930s[27] and during the 1960 to 1980s[28] with developments in human ecology.[29] An enduring link between the two disciplines is the influence cultural anthropology has had on cultural geographers. As Williams (1983: 4) notes, Sauer's work in particular borrowed ideas and concepts from the cultural anthropologists with whom he had close links at Berkeley.

5. Methods

Geography is based on two very simple but powerful interrelated contentions: that human and environmental issues are linked and cannot be separated; and that to understand these relationships, it is essential to consider issues of scale and spatial distribution.

Mapping is a crucial tool in understanding these relationships (see also chapter 10 on spatial modelling). Geographers use maps to both do research and to present their findings. Maps provide both powerful analytical and presentation tools. Maps can take many forms. Different cultures have developed a wide range of ways of mapping their lands and resources. In recent years the digital revolution has resulted in the development of powerful computer-based mapping systems. Geographers trained in geographic information systems (GIS) are able to superimpose numerous layers of biophysical and social data. Such sophisticated mapping techniques provide powerful new ways of exploring geography's tried and true interest in 'who gets what where'. Such technology has, for example, been used in recent years to explore issues of racism, highlighting for instance how 'communities of color' in the USA bear the brunt of poor environmental conditions (Pulido 2000). Likewise maps of crimes against women and maps of perceived safe and dangerous places provide powerful examples of the gendered nature of crime and safety.

Scale is a central tenet of geography and an understanding of it is crucial for addressing environmental issues (this is discussed further in chapter 12). Whether it is developing a response to an oil spill, designing the size of reserves required for effective conservation of endangered species, or implementing a community environmental education program, a policy is not going to work unless it is aimed at the right scale. Another example of the importance of scale in our increasingly globalised world is the connections between global, national, regional and local issues. Geographers have long had an interest in such relationships and the discipline provides powerful insight into handling such complex nested relationships.

Case study: the explanatory power of maps

Mapping can be a powerful analytical tool that provides insights that other research methods often fail to reveal. For instance, mapping I carried out as part of my research with an Indigenous Australian group has revealed a number of things that previous researchers in the area had failed to realise (Baker 1990, 1999). It showed that Yanyuwa men and women have quite different perspectives on the past. As part of a study of Yanyuwa responses to the arrival of Europeans on their land, I travelled with Yanyuwa people to now abandoned campsites. In an attempt to understand how various Yanyuwa perceive their land, I asked them to draw maps of how they remembered these old campsites being used. Yanyuwa men and women drew quite different maps (figures 11.1 and 11.2).

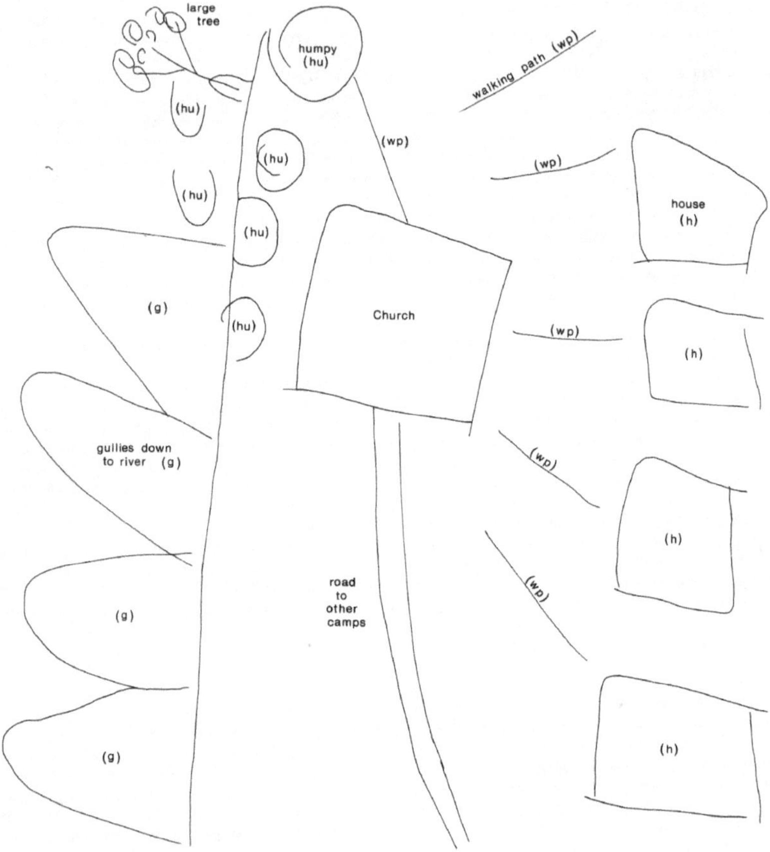

Figure 11.1. Bella Marrajabu's map.

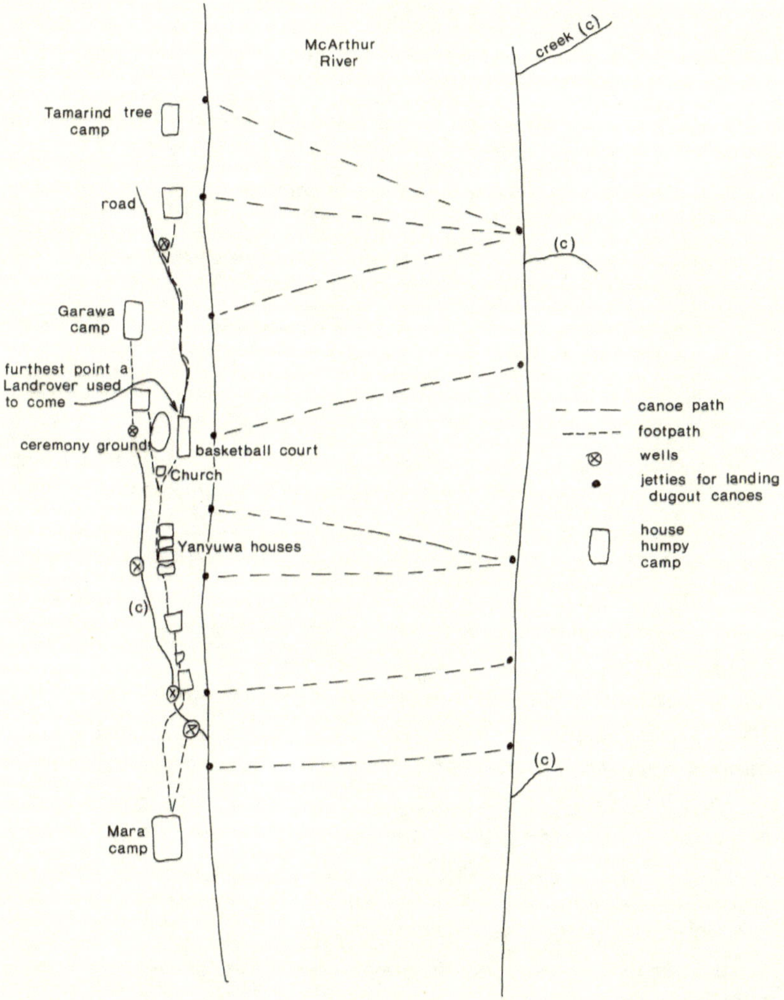

Figure 11.2. Andrew Dodd's map.

The reasons for the gender specific maps became clear when I questioned my Yanyuwa friends about their lives. Yanyuwa men and women, particularly in the 1950s, had very different personal geographies. Most Yanyuwa men went to work on the 'white side' (across the river) each weekday but would return each night to the camp. By contrast, the lives of women focused very much on the camp. Bella Marrajabu's map (like those of many other women) focused on the centre of her life, the church. Andrew Dodd's map (like those of many other men) focused on the river that was the access route both to his marine fishing and hunting grounds down stream and to the 'white side' where many men had jobs.

Another example of the power of mapping comes from the Yanyuwa transition from traditional 'bush' life to living in European established towns. Various historians have debated *why* Indigenous people made this transition at different times and places. By mapping the lives of Yanyuwa people it became possible to document *how* this process occurred in the Yanyuwa case. Because place of birth is of great religious importance to the Yanyuwa, like many other Indigenous peoples, everyone knew where several generations of their ancestors were born, and from this it was easy to construct a historical birth site map. Mapping the major birth sites of each of the four Yanyuwa clans reveals that traditional Yanyuwa society involved a seasonal round that involved each clan spending a few months each year at a number of favourite campsites.

An intriguing pattern emerged. When 'bush' births (i.e. occurrences where women gave birth in traditional circumstances) were mapped against the location and date of first European settlements in the area, it could be seen that European settlements in the area were at locations that were already the focus of Yanyuwa life.[30] Such a finding completely turns on its head the more usual question asked, 'Why did Aboriginal people move into European settlements?' Initially at least, the Yanyuwa did not move into European settlements at all: Europeans established their settlements at existing major Yanyuwa campsites.

6. Geography's relevance in addressing contemporary environmental problems

The relevance of geography to emerging global environmental issues has long been acknowledged. A 1973 US study on the importance of geographers in environmental issues (Lowenthal and Cooley 1973:

42) concluded the geography was particularly well placed to make a contribution to such issues for a wide range of reasons. These included geography's ability:

- to synthesise material that was usually the preserve of a wide range of other disciplines;
- to consider a 'complexity of explanation'; and
- to handle a range of information sources.

The relevance of geography to environmental issues has been greatly enhanced since the early 1970s by the rapid globalisation of such issues. As Liverman (1999: 108) noted, geography is of enormous value to global change research because it links environmental and social sciences and is sensitive 'to processes occurring at different scales, and the ability to provide conceptualized and comparative case studies'.

Geography is an unusual discipline in the way it combines social and biophysical sciences. Geography is about synthesising material from a range of sources. Geographers need to know about a range of fields and be capable of working effectively with experts in other disciplines wider afield. Such interdisciplinary skills are crucial in dealing with contemporary human–environmental problems and in dealing with complex and multi-faceted human–environment issues. Likewise, the appreciation geographers have of the complexity of causation is important because 'quick fixes' to environmental problems often fail. Often, so-called 'solutions' to environmental problems fail because they have not understood the complex interplay of environmental and cultural factors. Engineering solutions to soil erosion, for example, might not address the underlying social factors that are the root cause of soil erosion (see also chapters 8 and 9). If social factors such as insecure land tenure, cost-price squeezes or declining balances of trade are the main factors causing farmers to farm in unsustainable ways, engineering solutions alone are likely to fail.

7. Conclusions

Geographers have valuable skills to bring to the study of human–environment issues. Of crucial importance is our capacity to integrate social and biophysical issues and take into consideration the spatial dimension of these highly spatial issues. Our ability to integrate findings from other disciplines and to work effectively with researchers from a wide range of areas is also a part of the interdisciplinary 'glue' that is required in the crucial task of more fully understanding human–environment relationships.

8. Further reading

The following references reflect a range of views on the place of the discipline of geography in understanding the environment: Croft 1999, Knox and Marston 2001, Liverman 1999, Lowenthal and Cooley 1973, Meinig 1986, Sauer 1925, 1963, Stoddart 1986, Williams 1983.

There are a huge range of geographic journals published in a wide range of languages around the world. Some useful and accessible journals include:

Annals of the Association of American Geographers
Applied Geography
Area, Journal of the Institute of British Geographers
Australian Geographic Studies
Australian Geographer
Canadian Geographer
Environment and Planning
Professional Geographer
Progress in Human Geography
Progress in Physical Geography
Singapore Journal of Tropical Geography
Transactions of the Institute of British Geographers

An enormous number of geography resources can be found on the web using any search engine — some good sites to start at are:

http://sres.anu.edu.au/people/richard_baker/index.html — author's webpage with many geography links
http://www.chelt.ac.uk/gdn/ — geography discipline network
http://www.nationalgeographic.com/ — National Geographic Society
http://uweb.txstate.edu/~ms32/CGGE/index.html — the Online Center for Global Geography Education

http://www.canadainfolink.ca/geog.htm – resources for teaching and learning about Canadian Geography
http://www.geoexplorer.co.uk/index.html – UK geography portal

12 BRIDGING THE DIVIDES

R. Quentin Grafton and
Libby Robin

1. Introduction

Comparing and contrasting some of the key ideas discussed in the previous chapters, it is clear that understanding the environment may be limited by the way different domains of knowledge use language, even plain language. A word like 'environmental' adds humans to earth science (chapter 8), but nature to history (chapter 2). The context for the word completely alters its meaning and emphasis. 'Environmental' is also contested because it is a 'good' (morally virtuous) word. Engineers sometimes use it to make technical solutions appear more socially acceptable (chapter 9). Words are defined differently in different disciplines and the disciplinary backgrounds of researchers (consciously or unconsciously) inform and frame the questions considered or overlooked. While methodological approaches tend to have disciplinary origins, research questions will also sometimes be shaped by world events beyond any disciplinary frameworks, and different answers will be found in different eras because of that context.

In this chapter we bring together some of the insights generated by considering ten of the domains that contribute to 'environmental studies', particularly in relation to *communication* and *connection*. We open with discussion of 'origins' that examines some of the events beyond scholarship that have shaped environmental studies, driving a need to reach beyond single disciplines. The next section considers the language of environmental scholarship and how several key 'bridging' concepts are defined across different domains, and how to improve communication between domains using the concepts. We also reconsider the issues of disciplinary distance introduced in the first chapter, and explore some of the differences between interdisciplinary and disciplinary approaches, and between theoretical and empirical thinking (within and across disciplines) that can interfere with building ideas that connect across disciplinary worlds. We conclude with a reflection on the need for both connection and disciplinary difference if we are to begin to understand the complex and contested idea of 'environment'.

2. Origins of environmental studies

Like each of the disciplines and domains considered in the preceding chapters, the idea of a holistic, interdisciplinary 'environmental studies' has its own history. Interdisciplinarity has been more or less fashionable at different times, depending on what is considered to be the ideal or preferred forms of scholarly endeavour. It is helpful to understand the context in which environmental studies has grown most recently, and to recognise that other eras have also valued interdisciplinarity. What are the trends that favour mixing ideas and methods from different disciplines, or that tend to reinforce the separation of disciplines, and sometimes even subdisciplines from each other? The brief sketch below does not exhaust the possibilities, but it does suggest some events in the world beyond scholarship that have actually changed the direction of environmental studies.

Like many aspects of our disciplines, the interdisciplinary environmental studies that inform scholarship today emerged in the 'revolutionary' 1960s, but they also build on earlier revolutionary philosophies begun in the mid-18th century. The Enlightenment emphasised original research, creating new knowledge, pushing further into new domains, especially the ones we now refer to as scientific. In the 1960s, there was another shift away from classical thinking. Knowledge was not about rehearsing what others had learned before. Interconnections between disciplines became important nodes for new knowledge. Disciplinary categories forged in traditional 19th century German-style universities were seen to be barriers, not helpful to dealing with the complex issues of environment so apparent in the 'crisis' of the 1960s and 1970s.[31] Ecology, for example, began to consider not just an animal or plant and its *biological* place in 'the web of life', but the physics of organisms and energy flows in their environment. The economy of the ecosystem was informed by understandings from both *economics* and *physics* in closed total system models based on energy flow (see chapter 7).

Science combined with social and historical factors to create the background for the 1960s political revolution. The atomic age of the 1940s and 1950s introduced the possibility of literally blowing up the whole world. Hiroshima in Japan and many of the nuclear testing sites in the Pacific (Bikini Atoll, Mururoa Atoll) gave glimpses of Earth transformed into moonscape. The world's biological support systems were not only compromised by the latest devices of atomic physics, they were also under 'attack' on a range of fronts by the voracious development of the era following World War II. For instance, the 1950s marked the beginning of unprecedented economic growth, terrestrial species loss and population expansion (Steffen et al. 2004: 4–6). These developments were increasingly global, affecting not just the places in the West where the money was made, but also poor nations and non-Western communities everywhere.

The idea of *limits* and of the finite resources of Earth emerged in literature, film, music and popular culture, as well as in the scholarly disciplines. There was an increasingly angry sense that the *world was finite*, and a growing fascination with ideas that suddenly seemed to be everywhere. Of course some images *were* everywhere! In July 1969, the picture of the floating blue earth as seen from the Apollo 11 moon mission appeared in just about every media outlet in the Western world. It captured succinctly the notion that *there is only one Earth*.[32] 'Spaceship earth' was another metaphor from this era, used by Kenneth Boulding (1966) to present the notion of a finite (not indefinitely growing) economic system.

Environmental studies are not, however, merely 'a child of environmentalism'. The mid-19th century scholar was often a polymath (or multidisciplinary expert) with a passion for nature study. People we would call 'scientists' generally referred to themselves as 'natural philosophers' until well into the 19th century. Some modern origins of interdisciplinary environmental studies can be traced to 19th century British naturalists, such as the geologist Charles Lyell or the biologist Charles Darwin. But the professionalisation of most disciplines in the late 19th century stifled this early flowering, and delayed many interdisciplinary connections. Moreover, the political impetus that establishes a discipline or interdiscipline at a particular time (or consigns it to oblivion) is not the same as the scholarly questions that sustain it.

Much vibrant interest in the environment has its roots in environmental passion rather than logic. Passion is a strength, but is also a weakness in scholarship. It can make for lively prose, but it is important to know at all times that we have *evidence* for our arguments. We need to acknowledge the value of the broadest possible range of evidence (including non-Western understandings well beyond the domains of a Western-style university), whilst maintaining a very clear understanding of what arguments such evidence can support and challenge. This requires an understanding of the context of the evidence and appropriate techniques for interpreting it.

3. Which domains?

One of the most debated areas is which disciplines and domains should be invoked in 'environmental studies'. A case can be made for every discipline — but choosing the right ones for any given issue requires judgement.

To take a simple example, 'Which disciplines should be represented in this book?' was a question hotly debated by our authors and reviewers. Some patterns emerged in this context: the champions for missing disciplines generally came from adjacent ones, familiar with the style and useful concepts in nearby areas. For example, the policy and institutional analyst could not describe his domain without a good understanding of government and legal systems, and so felt political theory and law to be glaring omissions. An ecologist could not conceive of talking about biodiversity conservation without a good geneticist in the team, and she therefore felt that a book about the environment — if it were not to be simply an 'earth/air/water' view of the environment — needed a chapter on genetics. One historian was concerned about the lack of a philosopher, since eco-philosophies are absolutely essential to understanding the potential historical agency of non-human actors, such as other species or physical forces. Just about every politically important environmental issue — climate change, soil acidification, water quality — all demand a knowledge of chemistry, yet we have no chapter devoted to chemistry. This book, like proposals for solutions to many environmental problems, cannot in the end go on forever. While we acknowledge gaps in the domains covered within the book, our purpose is not to be comprehensive (even if we could be), but to *bridge* the gaps across knowledge domains where they exist. This can be achieved whether we cover ten or a hundred disciplines or domains although there is clearly a cognitive limit on the number of interconnections that can be usefully made by researchers (and by you, the reader).

The ten domains we have represented here offer windows onto a variety of styles, world views, methods and possible ways of conceiving the environment. The fact that there is a workable number of them is important to the possibilities of connection. Our authors come to environmental studies from such different backgrounds that they offer great scope to contest many of the ideas held up as sacrosanct in each other's training. Such contestation is the beginning of connection, of getting beyond that original disciplinary training. And these ten domains are not just a hypothetical. They really are working together in one place, at the Australian National University in 2004.[33] This is the strength and context for, but also the limitation of the choices, made in the book. The human capacity for interconnections is rich, but not infinite. Relationships between people and their tolerance for a diversity of viewpoints, may, in practice, be more important than their disciplinary representativeness.

4. Contested concepts

This section focuses on four major concepts crucial to environmental studies that have emerged in the various chapters in this book. What do they mean, when we start to bridge the divides? How can meanings draw on different disciplines and domains? If key words mean different things in different contexts, this adds difficulty to the task of working across disciplines. Clarity of concepts is essential (and much contested!) in developing research agendas with people trained in other disciplines or world views. The art of choosing words that reach beyond disciplinary training to build new insights is challenging — and often much harder than travelling the well furrowed roads of standard disciplinary concepts, examples and shorthands. The four concepts we consider here are *scale, systems, landscape* and *sustainability*. They are not the only ones, but serve as good examples because they are broad, and emerge from many or even most environmental domains. The fact that they are 'contested' is not in itself a bad thing. Being aware of difference and working through such contestation is a way of clarifying meaning and forging new understandings.

Scale

In studies of the environment, scale frequently takes on two different dimensions, one relating to time and the other to space. The other way scale can apply is in relation to the level of detail or aggregation adopted in the analysis. Timescales range from the short periods with which hydrologists (chapter 9) sometimes work — as fine as a few hours or less — to those of earth scientists who can examine the environment over time scales of millions of years (chapter 8). Policy-makers may work with time scales as little as a few weeks or years (chapter 6) and their decisions are frequently shaped by the political term of government, or the constraints of the financial year. On the other hand the environmental earth sciences work to understand change generated by physical 'drivers', such as continental drift, that may be virtually impossible to detect over a human lifetime, but make great changes over thousands or millions of years. Spatial scales are of particular concern to geographers, and choosing the right one for the task at hand is matter of considerable skill (chapter 11). Global effects may not be important locally, and vice versa. It depends on the question.

Methods used to study environmental change at different time or spatial scales are radically different, and demand very different levels of *detail* of analysis. For instance, economists divide their discipline into two main parts: macroeconomics, the study of economic aggregates such as the level of economic activity at a national level, and microeconomics, the study of economic behaviour at the level of the individual (see chapter 4). Even the notion of *habitat*, the place where a particular organism lives, can vary enormously in terms of scale (chapter 7). A habitat may be as large as the Amazon rainforest for some species of birds, or as small as a tiny island for a flightless bird. Getting the appropriate scale of analysis for the problem is a first step to a better understanding of the environment, and will influence the choice of method or disciplinary style applied to the problem.

Systems

A feedback system may be defined as a set of entities that interact to constrain each others' behaviour.[34] It is a self-contained whole and its dynamics describe the way its state variables respond to endogenous (arising inside the system) and exogenous (arising outside the system) forces. The network of feedback loops between variables describes the structure of a system and time-dependent changes in the values of its state variables represent a system's behaviour. A system's boundary encloses the endogenous variables that influence each other via feedback loops. The endogenous variables, together with exogenous variables that either influence or force the system, need to be enclosed in any explicit model of the system. A system's boundary can be a physical construct as small as a Petri dish or as large as a solar system, but the boundary can also be a mental construct whereby the person modelling a system makes explicit decisions about what variables should be included and excluded from a model (Sterman 2000).

The concept of a system has been developed in parallel in a wide range of disciplines (some of which are represented in this book) which include anthropology, psychology, sociology, management, linguistics, economics, biology, geomorphology, engineering, physics, and mathematics. The relevance of feedback systems to environmental studies is that most natural and human processes are influenced by feedbacks. A full understanding of the structure and behaviour of these processes therefore requires a systems way of thinking. For example, many environmental processes (such as climate change) involve feedback causality whereby a variation in some factor A (the use of fossil fuels) causes a change in another factor B (the concentration of greenhouse gases in the atmosphere), which ultimately leads to a change in factor C (the Earth's surface temperature) that then feeds back (policy interventions, such as the Kyoto Protocol) to increase or decrease the value of factor A, and so on. Systems thinking, however, is much more than a method to better understand dynamic feedback processes. Just as we bridge the disciplinary divides in this chapter, so a systems way of thinking connects key variables and feedback processes to obtain a more holistic and clearer understanding of the environment.

Landscape

Landscape is a concept that works on one particular scale: it is commonly defined as an expanse of scenery that can be seen in a single view. It is generally continuous, and in some sense, a holistic place. But landscapes are also complex, and seen differently by different disciplines. The properties of a landscape are linked to and represented by *human* observation. For instance, landscape painting and landscape architecture emphasise the human creative effort. In science, the observer is less prominent. Whereas landscape architecture is focused on enhancing the intersection between the social and the physical in a given space, landscape ecology sees a landscape as a space defined by biological and physical properties. Landscape ecologists see landscapes as systems, particularly systems in (or out of) balance. Landscape ecologists value not just continuities, but also patchiness. Patchiness, a function of soil type, climate and environmental history, is a systemic way of studying how different adjacent ecosystems work together to support different species of plants and animals (biodiversity) (Ludwig and Tongway 1999).

Landscape is not just empty space; it is a *place* valued by humans or animals or plants (or all of these). The perspective of the viewer makes a great deal of difference to how much of the space is considered part of the landscape (the scale of the whole landscape). 'Sense of place' is another contested notion that links research in a wide range of social and natural sciences and the humanities. Place and landscape overlap with ideas of *home, location, belonging, country, orientation, borders* and *indigeneity* (Malpas 1999). All of these are dependent on who is doing the observing — or making the judgements. Scale may also play a part. A decision about whether a plant is (or is not) regarded as indigenous to a landscape may depend on the extensiveness of an area included by the landscape, and how long ago the plant or animal or its predecessors arrived from somewhere else.

Sustainability

Sustainability has become a key concept for most environmental studies, particularly since it has become entrenched in international political conventions after the Rio conference in 1992. Dovers (chapter 6) defines sustainability as a desirable, but not necessarily attainable, social goal on par with peace and democracy. Politicians may use sustainability in a deliberately vague way to signal to voters that their policies are good for the environment. Oil and forest companies may use sustainability in their advertising to give themselves a green image, even when others argue that their practices are anything but sustainable.

Various disciplines have defined sustainability in their own ways, but that is not inconsistent with its notion as a social goal. For instance, economists have defined sustainability as (a) the principle that present actions should not make future generations worse off than the current generation, and (b) the principle that the 'natural capital stock' or the overall quality of the environment and level of natural resources should be left undiminished over time. Both definitions incorporate the concept of a timescale that spans generations and reflect societal goals concerning the effect our actions today have on future prosperity and on our environment. Thus sustainability provides an objective, albeit contested, view about what we should do today while explicitly considering the consequences for our descendants and the state of the planet.

Connecting concepts

Good communication requires that audiences understand what such major concepts mean, particularly if we want to engage support for particular environmental outcomes. If we are working together to generate research we need to understand each other. We are also obliged to help ensure that our findings arrive in a coherent form, and, if they are relevant to the wider community, that they are accessible and clear beyond disciplines and interdisciplines. Communicating to a wider public, however, may be viewed by some of our narrower colleagues as a worthless or even damning activity. It may have costs. Bureaucratic assessments of academic output often discourage or discount popular writing and public speaking. But the popular is particularly important in environmental studies because it has such strong connections with community, and is of much interest beyond the scholarly world.

Communication devices may vary to suit different purposes. They might be as simple as writing a letter to the editor of a newspaper or developing a user-friendly website or something more involved such as writing a popular book. Not every scholar can write a bestseller about environmental issues, nor can every good writer speak well on radio. The point is that the chosen mode of communication should target a specific audience and different forms of communication are required for different audiences. For instance, a scholarly letter in the prestigious journal *Nature* will better target a scientific audience, but plain language and short messages are essential for media 'grabs' and for providing simple clear recommendations to policy-makers at meetings or public forums. When reaching out to different audiences the evidence of arguments given in a more popular form may need to be expanded upon in more nuanced writing, designed for an audience prepared to put in more time to follow up the initial message.

Connecting ideas from different domains demands a tolerance of how people trained in other ways might understand such concepts differently. Connections are not always transformative. Integration is not the only possible way of joining disciplinary forces and different knowledge bases. There are some situations where it is ethically important that difference be maintained. For example, in a national park, where Indigenous and Western managers work together, it is desirable to co-manage the land: to respect each other's ways of knowing about it and engage in active dialogue, but not to supplant one set of knowledge with another.

5. Disciplinary distance and interdisciplinarity

As foreshadowed in chapter 1, this book has grappled with disciplinary distance — the intellectual barriers (in time and effort) that hinder interdisciplinary communication. These barriers include, but are not limited to:

- language and technical terminology (key concepts)
- use and interpretation of mathematical constructs
- use and interpretation of story or narrative (especially where these emerge from non-Western contexts)
- understandings built up at very different scales, and
- use and interpretation of technology.

Major differences also exist across disciplinary domains. For instance, economists tend to think that economic processes are part of the solution for overcoming environmental problems, while biologists and anthropologists may view such processes as a cause rather than a cure for such problems. Distance may also be described by the nature of the problems studied by its practitioners. For example, disciplines such as ecology (chapter 7), earth sciences (chapter 8), and geography (chapter 11) are particularly spatially sensitive, in that the questions that they seek to address are intimately related to particular landforms or landscapes. By contrast, the questions that economists or historians address often lack a spatial dimension — and may evolve differently if asked 'spatially', that is, how they apply in a precise location rather than as a general principle or over a time period.

By providing a concise introduction to ten different domains we have offered some insights to start building some of the 'bridges' that we need to better understand the environment. Understanding the way concepts are defined and framed by other domains is a first step. Another way forward is to understand and value alternative methods. Even something as seemingly universal as the scientific method may be defined in practice very differently by natural scientists and social scientists, and by different natural scientists (for example, physical and biological). It is even contested within disciplines.

As we move further away from our own discipline or subdiscipline, the cost in time or intellectual effort increases. This is likely to further increase *at an increasing rate* as we cross out of similar areas of study towards unrelated domains of study. For example, if we are trained in physics, other physical sciences such as chemistry or earth science will be less different from our comfort zone than economics, history or other social sciences. The cost of reaching across the disciplinary divide is significantly greater, as the difference is not just in methods, but also in allowable questions. This book has sought to reduce some of the costs of cross-disciplinary communication or disciplinary distance, in the interests of promoting a more diverse and richer approach to understanding the environment, to get a better sense of 'the whole elephant' (as described in chapter 1).

The cross-fertilisation of ideas from different knowledge domains is called *interdisciplinary* or cross-disciplinary research and is practised, in various forms, with respect to a whole range of environmental problems. Some researchers want to push this further and transcend all existing disciplinary domains in *transdisciplinary* research. Connections between multiple disciplines do have the potential to create new ways of thinking, just as new subdisciplines can sometimes force parent disciplines to reinvent themselves. Establishing links across knowledge domains is clearly important for a better comprehension of many environmental problems, but interdisciplinarity does not automatically create a better understanding of the environment. Just as there are limits to any single discipline's comprehension of the world around us, so too there are limits to what an interdisciplinary understanding can give us.

The foregoing chapters collectively suggest that environmental studies must begin in the real world. They must consider carefully the environmental problem to be tackled. The environmental question dictates the boundaries (the limits and opportunities) for an interdisciplinary approach. For example, if we wish to predict the global consequences of climate change over the next 100 years then we would need to try to understand a huge range of interactions and feedbacks (social, economic, physical and biological) that demand a great deal of interdisciplinary research on a large scale. By contrast, it would be wasted effort to use this large-team interdisciplinary approach in considering the short-term consequences to a particular frog population of the construction of a small farm dam. Indeed, trying to obtain a detailed understanding of all possible social, economic, physical and biological connections of the new dam on the frog population, may render the problem so complicated that the farmer cannot wait for our deliberations before building the dam. If we simply concentrate our research efforts on the ecological processes in this case, we can probably explain enough of the variables to reach a reasonable decision about where best to locate the dam.

Whatever the environmental problem, we must recognise (and acknowledge) boundaries on what can realistically be considered. These boundaries are not necessarily disciplinary boundaries, but are limits devised to make environmental problems understandable, and to scope research. Boundaries are also influenced by spatial and temporal dimensions. For example, in the case of the effects of a small farm dam on a particular frog population this year, both the temporal and spatial dimensions are relatively limited. Thus, it may be better to go deep in terms of our domain or discipline by focusing our attention on the ecological processes. On the other hand, with problems such as global climate change that have a huge spatial dimension and cover many decades, or even centuries, we may need to be broad (or much more interdisciplinary) to account for important feedbacks and connections that transcend disciplinary boundaries. Neither disciplinary nor interdisciplinary approaches are best in every case. It depends on the problem at hand.

Figure 12.1 depicts the concept of disciplinary and interdisciplinary approaches to understanding the environment. The double-headed arrow represents a series of points, or a domain, along which we may wish to be, depending on the nature of the problem that we are studying. At the one extreme we find interdisciplinary thinking and, at the other, disciplinary thinking with a myriad of combinations in between. Given the limitations of the human mind, it makes no sense to make a 'Holy Grail' with our thinking (or any approach to which we might aspire, but can never reach). No one individual can have a deep understanding of *every* discipline. Even a cross-disciplinary group cannot know everything because of barriers to communication, and there are always more disciplinary perspectives possible than are represented by any group. Moreover, even if it were achievable in a collaborative process, the costs (time and effort in collaboration) in reaching the 'Holy Grail' are likely to be so high as to render the net benefits negative. The point is not that interdisciplinary collaboration is not worthwhile – only that, at some point, boundaries must be placed on the extent of the disciplines spanned, to make a problem tractable.

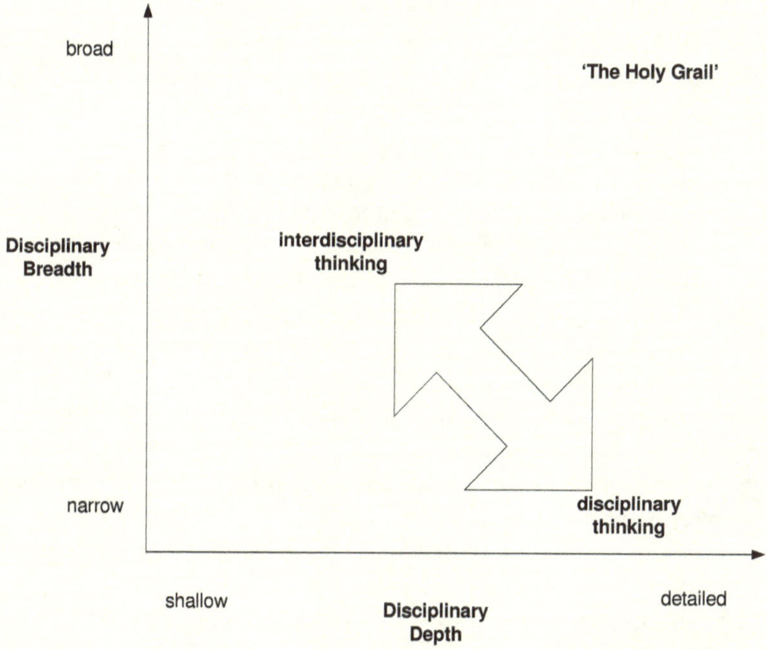

Figure 12.1. Disciplinary and interdisciplinary thinking and the 'Holy Grail'.[35]

6. Theory and observation

Whatever the domain of knowledge, understanding requires both mental models (theories) of processes and outcomes and methods to test these models. Even within disciplines these two approaches are often separated, in the sense that many theoreticians often do little if any empirical testing, while empiricists often test the theories or models developed by others. This is not necessarily a problem in the sense that researchers should specialise at what they do best, but can lead to a divide between the empirical testing and observations and the explanations for the causes of observed phenomena.

Sometimes observation will be ahead of the theory when measured phenomena are incompatible with existing models, and this triggers new theories. Once a new theory is put forward, it will be questioned and tested by empiricists. The push and pull between the two is essential to both. Sometimes theories generated by one discipline will inform practice in another, which makes the theoretical–empirical conversation even more complex. For example, in the 19th century, a number of different scholars observed an apparent matching of the shapes of the continents on either side of the Atlantic Ocean (in particular Africa and South America). In 1912 Alfred Wegener put forward a theory of continental drift (the idea that the continents shifted in relation to each other). Continental drift accounted for the matching continental shapes — and also some of the distribution of many of the world's biological species, but the theory's power was limited by its lack of a mechanism to explain the movement of continents. It attracted new interest in the early 1960s, when geologists developed theories of plate tectonics (or seafloor spreading), thus providing the crucial mechanism whereby the continents could move. The biogeographers interested in the distribution of species were quick to revisit the earlier theories, and species distribution is now commonly explained in terms of the movement of the Earth's plates.

A new theory can also make us observe the real world differently, and ask new questions. For example, one of the predictions of Einstein's theory of general relativity was that light would be bent by a greater extent if it went close to a very large gravitational field (such as a star) than if one accepted the Newtonian view of the universe. The prediction based on Einstein's theory could not be tested until a real world situation arose to observe its effects. Despite the fact that the measurements were imprecise and conflicting, observations of a solar eclipse in 1919 were interpreted to support Einstein's prediction from his general theory (Collins and Pinch 1993). Indeed, Britain's Astronomer Royal at the time went so far as to state that the observations confirmed Einstein's general theory, which, of course, is not the same as confirming the prediction from the theory!

The test of general relativity in 1919 is instructive as it shows that scientists chose to look for and then found evidence for light deflecting as predicted by Einstein. However, those undertaking the observations were also influenced by Einstein's theory in their interpretation of imperfect observations. In other words, the apparent belief in Einstein's general theory by those interpreting the data obtained from the solar eclipse led them to confirm that the data supported a prediction from the theory when, in fact, the empirical evidence provided, at best, mixed support for the prediction. Thus the notion of scientists (or social scientists) as unbiased observers and analysts is not always true — everyone is influenced by the prevailing wisdom and opinions of others and sees the world through a lens that may not be as dispassionate or scientific as they may like to believe.

Although the development of theory, models or mental maps and their subsequent testing is common to all domains of knowledge, how it is done varies enormously. The (idealised) physical sciences provide a model for a rigorous scientific method, but this should be adapted for most disciplines. The method involves the formulation of hypotheses that either explain or predict phenomena that are then tested to determine if they can be falsified or disproved with empirical measurement or observation. However, the failure to disprove a model or hypothesis does *not* make it true because empirical observations may also be consistent with alternative theories. Moreover, some theories that have subsequently been shown to be false, such as the classical Greek model (Ptolemaic system) of the solar system with the Earth at the centre, generated surprisingly accurate predictions of planetary motion. The same logic also holds when observations are inconsistent with theory — namely, it does not necessarily mean that the theory is false. This is because the failure of observations to agree with theoretical predictions may lie in improper observations or analysis of the observations.

The scientific method works best when as many variables as possible can be controlled for, and the observations or 'experiments' can be repeated. When considering phenomena that we can only imperfectly measure or observe (such as obtaining precise measurements of the effects of a solar eclipse), it is more difficult to prove or discount a theory. In this case, it is not clear what might be causing inconsistency — errors in the theory, errors in observation or measurement, or both. It follows, therefore, that a test of a theory or model is also a test of the method chosen to test theory. However, if differences between a theory's prediction and observations exist, the explanation for the inconsistency should be logical and as simple as possible. It should not be so vague or all encompassing as to make it impossible to prove the theory wrong!

It is significantly more difficult to falsify a theory in the real world than in a laboratory where variables can be manipulated systematically, one at a time. Many experiments cannot be repeated for ethical or practical reasons in the real world, making it difficult to falsify (or even modify) existing models or to choose between competing models. Such differences contribute to cross-disciplinary barriers. Researchers who use (or claim to use) the scientific method often feel uneasy when trying to understand theories (or paradigms) that cannot be tested in ways familiar to them. Similarly, researchers who do not use laboratories for theory testing may feel that a scientific approach constrains what can be studied, and what questions it is possible to ask, thus narrowing the scope of study and avoiding what, to them, seem critical issues. Nevertheless, most researchers would agree that models and theories should be tested against the real world, where possible, to verify whether theory and observation are consistent. But consistency may also be defined differently by different disciplines. For example, it may be shown by a chain of cause and effect mapped out logically through narrative, quantified through a spatial model or perhaps expressed through a systems analysis.

Models, theories and stories are simply, no more and no less, devices that we use to better understand the world around us. If they help us to predict events or outcomes, or give us insights into complex phenomena, then they are to be valued. If we view theory in this way, paradigms are simply constructs that help us to make sense of the world. Provided we do not use them as a means to close off new ideas and observations, they perform a useful purpose. For instance, what most people believed about the world two centuries ago would be ridiculed today, but by observation, better measurement, falsification of theories and the emergence of new paradigms we have developed new insights and understanding. Such is the sort of progress we might hope for the future.

7. Concluding remarks

Because the environment is a multi-faceted subject for study, it is unlikely that any one approach will throw up all the answers – or even the questions. Connecting the different scholarly ways of understanding the environment broadens the base on which questions are asked, and allows for a diversity of answers to those questions.

The environment embraces the largest possible physical systems and the most complex social organisations. Most of us, in some way, must deal with the interrelations between people and environments, and people are also best approached from a range of angles – what reaches one person or group will not necessarily touch another. The 'edge' between people and environments is often an interdisciplinary one.

Integration of knowledge across disciplines requires a matching of question to method. This does not always mean teamwork. Some interdisciplinary work will have a single author, whilst other works may be the seamless products of the collective consciousness of several authors. Possibly the greatest barrier to integration is a reluctance to question the questions: lateral solutions are often the product of lateral question-framing. The key steps in connecting our understanding are asking the right questions, comprehending the boundaries of the problem, establishing what can be done given constraints of time and effort, and then communicating the insights in an effective way to multiple audiences.

Acknowledgements

The authors are grateful for the helpful comments and suggestions of Rob Heinsohn and Barry Newell in the preparation of this chapter.

Endnotes

Chapter 1. Towards an understanding of the environment

1 W.D.L. Ride *A guide to the native mammals of Australia* (Oxford University Press, Melbourne 1970). The comment was a retrospective one made in 1999.

Chapter 2. History and the environment

2 John Mackenzie, Keynote address, University of Sussex conference on world environmental history, March 2003.

3 See, for example, Jared Diamond (1997) *Guns, germs and steel: the fates of human societies,* Jonathan Cape, London; Tim Flannery (1994) *The future eaters: an ecological history of the Australasian lands and people,* Reed Books, Chatswood, NSW; and Tim Flannery (2001) *The eternal frontier: an ecological history of North America and its peoples,* Text Publishing, Melbourne.

4 See also Vandana Shiva (1991) *The violence of the Green Revolution,* Zed Books, London.

5 The ratio between maximum and minimum annual flows was calculated at 4705.2 for the Darling and 15.5 for the Murray, compared with 1.9 for the Rhine in Switzerland and 2.4 for the White Nile in Sudan (FAO AquaStat Data, 2002).

6 http:www2.h-net.msu.edu/~environ/historiography/ustechnology.htm

7 Pyne (1995) is one of a number of his major works in his *Cycle of fire* series.

Chapter 9. Hydrology

8 Civil engineering in many areas has been rebadged as environmental engineering.

9 The transmission coefficient (hydraulic conductivity) of water through some soils can decrease by a factor of ten million as soil water content decreases by a factor of 20.

10 Rainfall, for example, is often measured in a 200 mm diameter raingauge.

11 Nathaniel Beardmore *Manual of hydrology, etc.,* London: Waterlow & Sons, 1862. In 1851 Beardmore also published *Hydraulic tables: properties and strength of materials, useful numbers and logarithms,* arguably an earlier text for hydraulic engineers.

12 Evapotranspiration involves evaporation from the soil and water bodies together with water transpired or evaporated from plant surfaces.

13 Initially dam theory involved the prediction of the size or operational rules required for a dam for a specified reliability of water supply given highly variable inflows and demands.

14 An isotope of an element is an atomic species that has the same number of protons in its nucleus as the element but a different atomic mass (different number of neutrons).

15 Nucleation particles are small particles of dust that allow water vapour to condense or nucleate.

16 Potential evaporation is the quantity of water evaporated per unit area per unit time from an idealised, extensive water surface under the existing atmospheric conditions. The supply of water to the surface does not limit evaporation.

17 For conservative substances like water the last term is not required.

Chapter 10. Mathematical and statistical sciences

18 E.T. Bell, 1952. *Mathematics, queen and servant of science.* G. Bell & Sons, London.

19 For more on this see, A.D. Aczel, 1999. *God's equation: Einstein, relativity and the expanding universe.* Four Walls Eight Windows, New York.

20 For more detail see M.F. Hutchinson (1998) Interpolation of rainfall data with thin plate splines: I. Two dimensional smoothing of data with short range correlation. *Journal of Geographic Information and Decision Analysis* 2(2): 153–167.

21 Earlier work in this field was undertaken by Krige in the 1950s to estimate the spatial distributions of ore bodies. The theory was largely established by Matheron in the 1960s. (See Chilès and Delfiner 1999 for more detail).

22 These include partial splines and additive tensor product spline models.

23 See Lumley, J.L. and Panofsky, H.A. 1964. *The structure of atmospheric turbulence.* John Wiley and Sons, New York.

Chapter 11. Geography

24 A detailed outline in English of Ratzel's ideas is that of E. Semple (1968) *Influences of geographic environment on the basis of Ratzel's system*

of anthropo-geography, Russel and Russel, New York, reissue of 1911 publication.

25 There is a strong parallel here with R.H. Tawney's idea that historians need strong boots (see chapter 2).

26 For a discussion of this movement see R. Baker (1997) Landcare: policy, practice and partnerships: *Australian Geographical Studies* 35(1): 61–73.

27 Examples include: Febvre, L. 1925. *A geographical introduction to history*. Kegan Paul, Trench, Trubner and Co., London. Forde, C.D. 1934. *Habitat, economy and society: a geographical introduction to ethnology*. Methuen, London.

28 Norwine, J. and Anderson, T.D. 1980. *Geography as Human Ecology*. University Press of America, Washington DC; Eyre, S.R. and Jones, G.R.J. 1966. *Geography as human ecology: methodology by example*. Edward Arnold, London are two good examples of this era.

29 Ellen, R. 1988. Persistence and change in the relationship between anthropology and human geography. *Progress in Human Geography*, 12: 229–262, especially p. 241.

30 For more details see Baker 1999.

Chapter 12. Bridging the divides

31 Environmental 'crisis' and 'revolution' were the rhetoric driving politics in the 1960s and 1970s, decades of high individualism. They greatly influenced the scholarly disciplines of this period. In the 1980s and 1990s, there was more influence on international and global solutions to environmental issues, and less emphasis on the crisis mode, and the scholarly disciplines and domains have since this period worked more towards systemic solutions and policy reform at all levels of government. 'Sustainability' is the rhetoric of this era. The interdisciplinarity forged in the earlier crisis era has continued beyond it.

32 This was the title of a well-known book by Barbara Ward and Rene Dubos, *Only one earth; the care and maintenance of a small planet* (1972: Norton, New York).

33 The authors trained in a wide range of universities all over the world, but they all work or have worked recently in the Australian National University's National Institute for Environment, either at the Centre for Resource and Environmental Studies (CRES), the

School of Resource, Environment and Society (SRES) or the National Centre for Epidemiology and Population Health (NCEPH).

34 This section draws heavily upon the work of Barry Newell who helped the authors define and describe systems.

35 Adapted from Grafton et al. 2004 figure 17.1 and Richmond 2001 figure 8-4, 114.

References

Ahern, M. and McMichael, A.J. 2002. Global environmental changes and human health. In Hester, R. and Harrison, R.M. (eds) *Global environmental change.* The Royal Society of Chemistry, Cambridge, pp. 139–160.

Allen, T.F.H. and Hoekstra, T.W. 1992. *Toward a unified ecology*. Columbia University Press, New York.

Anderies, J. 1998. Culture and human agro-ecosystem dynamics: the Tsembaga of New Guinea. *Journal of Theoretical Biology* 192(4): 515–530.

Baker, R. 1990. Coming in: Yanyuwa contact history. *Aboriginal History* 14: 25–60.

Baker, R. 1999. *Land is life: from bush to town — the story of the Yanyuwa people*. Allen and Unwin, Sydney.

Bateson, Gregory 1979. *Mind and nature: a necessary unity*. Fontana/Collins, Glasgow.

Beaglehole, R. (ed.) 2003. *Global public health: a new era.* Oxford University Press, Oxford.

Begon, M., Harper, J.L. and Townsend, C.R. 1996. *Ecology: individuals, populations, and communities*. Blackwell Science, Boston, 3rd edition.

Bennett, C.J. and Howlett, M. 1992. The lessons of learning: reconciling theories of policy learning and policy change. *Policy Sciences*. 25: 275–294.

Berkes, Fikret 1999. *Sacred ecology: traditional ecological knowledge and resource management*. Taylor & Francis, Philadelphia.

Berlin, Brent 1992. *Ethnobiological classification: principles of categorization of plants and animals in traditional societies*. Princeton University Press, Princeton.

Blaikie, P. 1985. *The political economy of soil erosion in developing countries.* Longman, Burnt Mill, Essex.

Blaug, Mark 1997. *Economic theory in retrospect*. Cambridge University Press, Cambridge UK, 5th edition.

Boas, Franz 1940. *Race, language and culture*. Macmillan, New York.

Bocking, Stephen 1997. *Ecologists and environmental politics: a history of contemporary ecology*. Yale University Press, New Haven.

Boulding, K. 1966. The economics of the coming Spaceship Earth. In H. Jarrett (ed.) *Environmental quality in a growing economy*, Resources for the Future/Johns Hopkins University Press, Baltimore, Maryland, pp. 3–14.

Bousquet, F., Lifran, R., Tidball, M., Thoyer, S., and Antona, M. 2001. Agent based modelling, game theory and natural resource management issues. *Journal of Artificial Societies and Social Simulation* 4(2), http://www.soc.surrey.ac.uk /JASSS/4/2/0.html.

Bowler, P.J. 1992. *The Fontana history of the environmental sciences*. Fontana Press, Hammersmith, London.

Braudel, Fernand 1949 (French), 1972 (English). *La Meditérranée et le monde Meditérranéen à l' époque de Philippe II*, Collins, Paris (in English: *The Mediterranean and the Mediterranean world in the age of Philip II*, translated by Sian Reynolds).

Bredehoeft, J.D. 1997. Water management in the United States — a democratic process (who are the managers?). In Buras, N. (ed.) *Reflections on hydrology: science and practice.* American Geophysical Union, Washington DC, pp. 36-61.

Bretz, J.H. 1928. The Channeled Scabland of eastern Washington. *Geographical Review* 18: 446–477.

Brooks, S. and Gagnon, A-G. (eds) 1990. *Social scientists, policy, and the state*. Praeger, New York.

Brunner, R.D. 1991. The policy movement as policy problem. *Policy Sciences* 24: 65–98.

Budyko, M.I. 1974. *Climate and life.* Academic Press, New York.

Buras, N. (ed.) 1997. *Reflections on hydrology: science and practice.* American Geophysical Union, Washington DC.

Burgman, M.A. and Lindenmayer, D.B. 1998. *Conservation biology for the Australian environment*. Surrey Beatty and Sons, Chipping Norton, Sydney.

Carson, R. 1964 (first published 1962). *Silent spring*. Fawcett Crest, New York.

Chilès, J. and Delfiner, P. 1999. *Geostatistics: modelling spatial uncertainty*. John Wiley and Sons, New York.

Chorley, R.J., Schumm, S.A. and Sugden, D.E. 1984. *Geomorphology*. Methuen, London.

Coates, D.R. (ed.) 1972. *Environmental geomorphology and landscape conservation.* Dowden, Hutchinson and Ross, Stroudsburg, Pennsylvania, 3 volumes.

Coates, D.R. (ed.) 1976. Urban geomorphology. *Geological Society of America Special Paper* 174.

Cohen, J. and Stewart, I. 1995. *The collapse of chaos: discovering simplicity in a complex world*. Penguin Books, New York.

Collins, H. and Pinch, T. 1993. *The Golem: what everyone should know about science*. Cambridge University Press, Cambridge.

Connell, Daniel (ed.) 2002. *Uncharted waters*. Murray Darling Basin Commission, Canberra.

Connor, R.D. and Dovers, S.R. (2004). *Institutional change for sustainable development*. Edward Elgar, Cheltenham.

Considine, M. 1994. *Public policy: a critical approach*. Macmillan, Melbourne.

Costanza, R., dArge, R., deGroot, R., Farber, S., Grasso, M., Hannon, B., Limburg, K., Naeem, S., O Neill, R.V., Paruelo, J., Raskin, R.G., Sutton, P. and vandenBelt, M.R. 1997. The value of the world's ecosystem services and natural capital. *Nature* 387 (6630): 253–260.

Cressie, N.A.C. 1991. *Statistics for spatial data*. John Wiley & Sons, New York.

Croft, R. 1999. Geography matters for the environment of the twenty-first century. *Geography* 84 (4): 345–353.

Cronon, William 1991. *Nature's metropolis: Chicago and the Great West*. W.W. Norton, New York.

Daneke, G.A. 1989. On paradigmatic progress in public policy and administration. *Policy Studies Journal* 17: 275–296.

Davie, T. 2003. *Fundamentals of hydrology*. Routledge, London.

Davies, G.L. 1968. *The earth in decay. a history of British geomorphology 1578–1878*. MacDonald Technical and Scientific, London.

Davis, G., Wanna, J., Warhurst, J. and Weller, P. 1993. *Public policy in Australia*. Allen and Unwin, Sydney, 2nd edition.

Davis, J.C. 1986. *Statistics and data analysis in geology*. John Wiley Sons, New York, second edition.

Dingman, S.L. 2002. *Physical hydrology*. Prentice Hall, New Jersey, 2nd edition.

Donahue, John and Johnston, Barbara (eds) 1998. *Water, culture, and power: local struggles in a global context*. Island Press, Washington DC.

Dooge, J.C.I. 1997. Scale problems in hydrology. In N. Burras (ed.) *Reflections on hydrology: science and practice*. American Geophysical Union, Washington DC, pp. 85–143.

Dovers, S. 1997. Sustainability: demands on policy. *Journal of Public Policy* 16: 303–318.

Dovers, S. 2003. A policy orientation as integrative strategy. In Dovers, S., Stern, D. and Young, M. (eds) *New dimensions in ecological economics*. Edward Elgar, Cheltenham.

Dovers S. (in press). *Environment and sustainability: a policy handbook.* Sydney: The Federation Press.

Dovers, Stephen (ed.) 2000. *Environmental history and policy: still settling Australia*. Oxford University Press, Melbourne.

Dovers, Stephen, Ruth Edgecombe and Bill Guest (eds) 2002. *South Africa's environmental history: cases and comparisons*. David Philip Publisher, Cape Town.

Dovers, S. and Wild River, S. (eds) 2003. *Managing Australia's environment*. Federation Press, Sydney.

Dryzek, J.S. 1997. *The politics of the earth: environmental discourses*. Oxford University Press, Oxford.

Eagleson, P.S. 2002. *Ecohydrology, Darwinian expression of vegetation form and function*. Cambridge University Press, Cambridge.

Elliott, P., Wakefield, J.C., Best, N.G. et al. (eds) 2000. *Spatial epidemiology.* Oxford University Press, Oxford.

Fairbanks, R.G., Evans, M.N., Rubenstone, J.L, Mortlock, R.A., Broad, K., Moore, M.D. and Charles, C.D. 1997. Evaluating climate indices and their geochemical proxies measured in corals. *Coral Reefs* 16: S93–100.

Falkenmark, M. 1997. Meeting water requirements of an expanding world population. *Phil. Trans. Roy. Soc. London, B.* 352, 929–936.

Feller, W. 1967. *An introduction to probability theory and its applications.* John Wiley and Sons, New York.

Finizio, N. and Ladas, G. 1982. *An introduction to differential equations.* Wadsworth Publishing Co, Belmont, California.

Franklin, J.F., Lindenmayer, D.B., MacMahon, J.A., McKee, A. Magnusson, J., Perry, D.A., Waide, R. and Foster, D.R. 2000. Threads of continuity: ecosystem disturbances, biological legacies and ecosystem recovery. *Conservation Biology in Practice* 1: 8–16.

Gadgil, Madhav and Ramachandra Guha 1992. *This fissured land: an ecological history of India*. Oxford University Press, Delhi.

Garrod, Guy and Kenneth G. Willis 1999. *Economic valuation of the environment: methods and case studies*. Edward Elgar, Cheltenham UK.

Ghassemi, F. and Brennan, D. 2000. *Resource profile subproject: summary report.* ACIAR Project. An evaluation of the sustainability of the farming systems in the brackish water region of the Mekong Delta. ACIAR, Canberra.

Ghassemi, F., Jakeman, A. and Nix, H.A. 1995. *Salinisation of land and water resources.* University of New South Wales Press, Sydney.

Glacken, Clarence 1967. *Traces on the Rhodian shore.* University of California Press, Berkeley.

Gleick, P.H. 1993. *Water in crisis: a guide to the world's fresh water resources.* Oxford University Press, New York.

Gleick, P.H. 1998. *The world's water: the biennial report on freshwater resources 1998–1999.* Island Press, Washington DC.

Graf, W.L. 1996. Geomorphology and policy for restoration of impounded American rivers. In Rhoads, B.L. and Thorn, C.E. (eds) *The scientific nature of geomorphology.* John Wiley and Sons, Chichester, New York, pp. 443–473.

Grafton, R.Q., Adamowicz, W.A., Dupont, D.P., Nelson, H.W., Hill, R. and Renzetti, S. 2004. *The economics of the environment and natural resources.* Basil Blackwell, Oxford.

Griffiths, Tom 1996. *Hunters and collectors: the antiquarian imagination in Australia.* Cambridge University Press, Cambridge.

Griffiths, Tom 2002. How many trees make a forest? Cultural debates about vegetation change in Australia. *Australian Journal of Botany* 50: 375–389.

Griffiths, Tom 2003. The nature of culture and the culture of nature. In Hsu-Ming Teo and Richard White (eds) *Cultural history in Australia.* UNSW Press, Sydney, pp. 67–80.

Griffiths, Tom and Libby Robin (eds) 1997. *Ecology and empire: environmental history of settler societies.* Keele University Press, Edinburgh.

Grinde, Donald and Johansen, Bruce 1995. *Ecocide of Native America: environmental destruction of Indian lands and peoples.* Clear Light Publishers, Santa Fe.

Grove, Richard H. 1995. *Green imperialism: colonial expansion, tropical island Edens, and the origins of environmentalism, 1600–1860.* Cambridge University Press, Cambridge.

Guha, Ramachandra and Martinez-Alier, J. 1997. *Varieties of environmentalism.* Earthscan, London.

Guha, Ranajit 1982. *Subaltern studies I: writings on South Asian history and societies*. Oxford University Press, Delhi.

Hancock, W.K. 1976. *Professing history*. Sydney University Press, Sydney.

Harvey, D. 1984. On the history and present condition of geography: an historical materialist manifesto. *The professional geographer* 36: 1–11.

Hem, J.D. 1992. *Study and interpretation of the chemical characteristics of natural waters.* US Geological Survey Water-Supply Paper 2254, US Government Printing Office, Washington, 3rd edition.

Hetzel, B.S. and Pandav, C.S. 1994. *SOS for a billion: the conquest of iodine deficiency disorders.* Oxford University Press, Bombay.

Hewawasam, T., von Blanckenburg, F., Schaller, M. and Kubilk, P. 2003. Increase of human over natural erosion rates in tropical highlands constrained by cosmogenic nuclides. *Geology* 31(7): 597–600.

Hogwood, B.W. and Gunn, L.A. 1984. *Policy analysis for the real world.* Oxford University Press, Oxford.

Hornborg, A. 2001. *The power of the machine: global inequalities of economy, technology, and environment.* Rowman and Littlefield, Altamira.

Howlett, M. and Ramesh, M. 2003. *Studying public policy: policy cycles and policy subsystems.* Oxford University Press, Toronto, 2nd edition.

Howser, P., Hutchinson, M.F., Viterbo, P., Hervé Douville, J. and Running , S.W. 2004. Terrestrial data assimilation. In Kabat, P. et al. (eds) *Vegetation, water, humans and the climate.* Global Change – the IGBP Series. Springler-Verlag, Berlin.

Humphries, C.J., Williams, P.H. and Vane-Wright, R.I. 1995. Measuring biodiversity value for conservation. *Annual Review of Ecology and Systematics* 26: 93–111.

Hutchinson, M.F. 1995. Stochastic space–time weather models from ground-based data. *Agricultural and Forest Meteorology* 73: 237–264.

Hutchinson, M.F. and Gessler, P.T. 1994. Splines – more than just a smooth interpolator. *Geodema* 62: 45–67.

Hutton, J. 1795. *Theory of the Earth, with proofs and illustrations*. Edinburgh, 2 volumes.

Huxley, M. 1999. If planning is anything, maybe it is geography. *Australian Planner* 36: 128–133.

Jones, H.L. 1917. Translation of Strabo. *The geography of Strabo.* Putnam and Sons, New York.

Kangle, R.P. 1965. *The Kautilya Arthasastra.* University of Bombay, 3 volumes.

Kaosa-ard M. 2003. Poverty and globalisation. In Kaosa-ard, M. and Dore, J. (eds), *Social challenges for the Mekong River.* White Lotus, Bangkok, pp. 405–440.

Keesing, Roger and Strathern, Andrew 1998. *Cultural anthropology in a contemporary perspective.* Harcourt Brace College Publishers, Forth Worth.

Kendall, M.G. and Stuart, A. 1979. *The advanced theory of statistics, vol. 2.* Charles Griffin & Co, London.

Kloeden, P.E. and Platen, E. 1995. *Numerical solution of stochastic differential equations.* Springer, Berlin.

Knox, P.L. and Marston, S.A. 2001. *Places and regions in global context: human geography.* Prentice Hall, New Jersey.

Kooi, H. and Beaumont, C. 1996. Large-scale geomorphology: classical concepts reconciled and integrated with contemporary ideas via a surface process model. *Journal of Geophysical Research* 101: 3361–3386.

Kreyszig, E. 1999. *Advanced engineering mathematics.* Wiley and Sons, New York.

Lakoff, G. and Núñez, R.E. 2000. *Where mathematics comes from.* Basic Books, New York.

Larsen, R.J. and Marx, M.L. 1986. *An introduction to mathematical statistics and its applications.* Prentice-Hall, Englewood Cliffs, New Jersey.

Lasswell, H. 1951. The policy orientation. In Lerner, D. and Lasswell, H. (eds) *The policy sciences: recent developments in scope and methods.* Standford University Press, Standford, pp. 3–15.

Lieber, M. 1994. *More than a living: fishing and the social order on a Polynesian atoll.* Westview Press, Boulder.

Limerick, Patricia 1987. *The legacy of conquest: the unbroken past of the American West.* Norton, New York.

Lindblom, C.E. 1959. The science of muddling through. *Public Administration Review* 19: 79-88.

Linder, S.H. and Peters, B.G. 1989. Instruments of government: perceptions and contexts. *Journal of Public Policy* 9: 35–58.

Lindgren, E. and Gustafson, R. 2001. Tick-borne encephalitis in Sweden and climate change. *Lancet* 358: 16–18.

Little, Paul 1999. Environments and environmentalisms in anthropological research: Facing a new millennium. *Annual Review of Anthropology* 28: 253–284.

Liverman, D.M. 1999. Geography and the global environment. *Annals of the Association of American Geographers* 89(1): 107–120.

Lowenthal, D. and Cooley, R. 1973. Report of the American Association of Geographers task force on environmental quality. *The Professional Geographer* 25: 39–46.

Lowenthal, David 2000. *George Perkins Marsh: prophet of conservation.* University of Washington Press, Seattle.

Ludwig, J.A. and Tongway, D.J. 1999. Viewing rangelands as landscape systems. In Arnalds, O. and Archer, S. (eds) *Rangelands desertification.* Kluwer, Dordrecht.

Macarthur, R.H. and Wilson, E.O. 1967. *The theory of island biogeography.* Princeton University Press, Princeton.

Maidment, D 1993. *Handbook of hydrology.* McGraw Hill Professional, New York.

Maidment, D.R. (ed.) 1992. *Handbook of hydrology.* McGraw Hill, New York.

Malpas, J. 1999. *Place and experience: a philosophical topography.* Cambridge University Press, Cambridge.

May, P. 1992. Policy learning and policy failure. *Journal of Public Policy* 12: 331–354.

Mazoor, E. 1997. *Chemicals and isotopic groundwater hydrology.* Marcel Dekker, New York.

McIntosh, R.P. 1985. *The background of ecology: concept and theory.* Cambridge University Press, Cambridge.

McMichael, A.J. 2002. The environment. In Detels, R., McEwen, J., Beaglehole, R. and Tanaka, H. (eds) 2002. *Oxford textbook of public health.* Oxford University Press, Oxford, pp. 195–215.

McMichael, A.J., Campbell-Lendrum, Ebi K. et al. (eds) 2003. *Climate change and human health.* World Health Organization, Geneva.

McMichael, A.J., Githeko, A., Akhtar, R. et al. 2001. Human health. In McCarthy, J.J., Canziani, O.F., Leary, N.A. et al. (eds) *Climate change 2001: impacts, adaptation and vulnerability.* Working Group II, Third Assessment Report of the Intergovernmental Panel on Climate Change. Cambridge University Press, Cambridge, pp. 451–485.

McNeely, J.A., Miller, K.R., Reid, W.V., Mittermeier, R.A. and Werner, T.B. 1990. *Conserving the world's biological diversity.* International Union for Conservation of Nature and Natural Resources, Gland, Switzerland; World Resources Institute, Conservation International, World Wildlife Fund-US and the World Bank, Washington DC.

Meinig, D.W. 1986. *The shaping of America: a geographical perspective on 500 years of history*. Volume 1: Atlantic America 1492–1800. Yale University Press, New Haven.

Merchant, Carolyn 1980. *The death of nature: women, ecology, and the scientific revolution*. Harper and Row, San Francisco.

Merritts, F., De Wet, A. and Menking, K. 1998. *Environmental geology: an earth system science approach*. Freeman, New York.

Messer, Ellen and Lambek, Michael (eds) 2001. *Ecology and the sacred: engaging the anthropology of Roy A. Rappaport*. University of Michigan Press, Ann Arbor.

Myllyntaus, Timo and Saikku, Mikko 2001. *Encountering the past in nature*. Ohio University Press, Athens Georgia (first edition 1999).

Nash, Roderick 1989. *The rights of nature: a history of environmental ethics*. University of Wisconsin Press, Madison.

Nazarea, Virginia (ed.) 1999. *Ethnoecology: situated knowledge/located lives*. University of Arizona Press, Tucson.

O'Loughlin, E.M. 1981. Saturated regions in catchments and their relation to soil and topographic factors. *J. Hydrology* 53: 220–246.

Page, Talbot 1977. *Conservation and efficiency: an approach to materials policy*. Johns Hopkins University Press, Baltimore.

Pascual, M., Rodo, X., Ellner, S. et al. 2000. Cholera dynamics and El Niño-Southern Oscillation. *Science* 289: 1766–1769.

Patz, J.A., Hulme, M., Rosenzweig, C. et al. 2002. Increasing incidence of malaria since 1970 parallels regional warming in East Africa. *Nature* 420: 627–628.

Pawson, Eric and Brooking, Tom (eds) 2002. *Environmental histories of New Zealand*, Melbourne: Oxford University Press.

Peters, R.H. 1991. *A critique for ecology*. Cambridge University Press, Cambridge.

Philip, J.R. 1975. Some remarks on science and catchment prediction. In Chapman, T.G. and Dunin, F.X. (eds), *Prediction in catchment hydrology*. Australian Academy of Science, Canberra, pp. 23–30.

Philip, J.R. 1995. Desperately seeking Darcy in Dijon. *Soil Sci. Soc. Amer. J.* 59: 319–324.

Philip, J.R. 1998. Physics, mathematics and the environment: the 1997 Priestley Lecture. *Aust. Met. Mag.* 47: 273–283.

Pickett, S.T.A., Kolasa, J. and Jones, C.G. 1994. *Ecological understanding.* Academic Press, San Diego.

Pigou, Arthur C. 1920. *The economics of welfare.* Macmillan, London, 4th edition published 1932.

Pulido, L. 2000. Rethinking environmental racism with white privilege and urban development in southern California: *Annals of the Association of American Geographers* 90: 12–40.

Pyne, Stephen J. 1995. *World fire: the culture of fire on earth.* Holt, New York.

Raats, P.A.C., Smiles, D.E. and Warrick, A.W. (eds) 2002. *Environmental mechanics: water mass and energy transfer in the biosphere.* Geophysical Monograph 129, American Geophysical Union, Washington DC, USA.

Rahman, M.M., Mandal, B.K., Chowdury, T.R., Sengupta, M.K., Chowdury, U.K., Lodh, D., Chanda, C.R., Basu, G.K., Mukherjee, S.C., Saba, K.C. and Chakraborti, D. 2003. Arsenic groundwater contamination and sufferings of people in North 24-Parganas, one of the nine arsenic affected Districts of West Bengal, India. *Journal of Environmental Science and Health* 38(1): 25–59.

Rappaport, R. 1968. *Pigs for the ancestors: ritual in the ecology of a New Guinea people.* Yale University Press, New Haven.

Rhoads, B.L. and Thorn, C.E. (eds) 1996. *The scientific nature of geomorphology.* John Wiley and Sons, Chichester, New York.

Richmond, B. 2001. *An introduction to systems thinking.* High Performance Systems Inc., Hanover, New Hampshire.

Robin, Libby 1998. *Defending the Little Desert: the rise of ecological consciousness in Australia.* Melbourne University Press, Melbourne.

Rohatgi, V.K. 1976. An introduction to probability theory and mathematical statistics. John Wiley and Sons, New York.

Rose, D. 2002. With Sharon D'Amico, Nancy Daiyi, Kathy Deveraux, Margy Daiyi, Linda Ford and April Bright. *Country of the heart: an Indigenous Australian homeland.* Aboriginal Studies Press, Canberra.

Ross, S.M. 1983. *Stochastic Processes.* John Wiley and Sons, New York.

Sauer, C.O. 1925. The morphology of landscape. *University of California Publications in Geography* 2: 19–53.

Sauer, C.O. 1941a. Foreword to historical geography. *Annals of the Association of American Geographers* 31(1, March): 1–24.

Sauer, C.O. 1941b. The personality of Mexico. *The Geographical Review* 31(3): 353–364.

Sauer, C.O. 1963. *Land and life: a selection from the writings of Carl Ortwin Sauer*. Leighly, J. (ed.), University of California Press, Berkeley.

Schimek, M.G. (ed.) 2000. *Smoothing and regression: approaches, computation and application*. John Wiley and Sons, New York.

Scott, Colin 1996. Science for the West, myth for the rest? The case of James Bay Cree knowledge construction. In Nader, L. (ed.) *Naked science, anthropological inquiry into boundaries, power and knowledge.* Routledge, New York, pp. 69–86.

Sheail, John 2002. *An environmental history of twentieth-century Britain.* Palgrave, Basingstoke.

Shortland, Michael 1993. *Science and nature: essays in the history of the environmental sciences*. BSHS Monograph, Oxford.

Skinner, B.J. and Porter, S.C. 1995. *The dynamic earth*. John Wiley, New York.

Slaper, H., Velder, G.J.M., Daniel, J.S. et al. 1996. Estimates of ozone depletion and skin cancer incidence to examine the Vienna Convention achievements. *Nature* 384: 256–258.

Smith, Adam 1776. *The wealth of nations*. The Modern Library, New York, 1938 edition.

Smith, Eric and Wishnie, Mark 2000. Conservation and subsistence in small-scale societies. *Annual Review of Anthropology* 29: 493–524.

Steffen, W., Sanderson, A., Tyson, P.D., Jäger, J., Matson, P.A., Moore, B. III, Oldfield, F., Richardson, K., Schellnhuber, H.J., Turner, B.L. II and Wasson, R.J. 2004. *Global change and the earth system: a planet under pressure*. Global change – the IGBP series; Springer, New York.

Sterman, J.D. 2000. *Business dynamics: systems thinking and modeling for a complex world*. Irwin McGraw-Hill, Boston.

Stewart, I. 1989. *Does God play dice?* Blackwell, Oxford.

Stoddart, D.R. 1986. *On geography*. Blackwell, Oxford.

Stumm, W. and Morgan, J.J. 1996. *Aquatic chemistry: chemical equilibria and rates in natural waters*. John Wiley and Sons, New York, 3rd edition.

Tansley, A.G. 1935. The use and abuse of vegetational concepts and terms. *Ecology* 16, 284–307.

Tietenberg, Thomas H. 2002. *Environmental and natural resource economics*. Addison-Wesley, New York, 6th edition.

Tong, S. and McMichael, A.J. 1999. The magnitude, persistence and public health significance of cognitive effects of environmental lead exposure in childhood. *J. Env. Medicine* 1: 103–110.

Torgerson, D. 1986. Between knowledge and politics: three faces of policy analysis. *Policy Sciences* 19: 33–59.

van den Bergh, Jeroen C.J.M. 1996. *Ecological economics and sustainable development: theory, methods, and applications*. Edward Elgar, Cheltenham UK.

Vitousek, P.M., Mooney, H.A., Lubchenco, J. and Melillo, J.M. 1997. Human domination of Earth's ecosystems. *Science* 277: 494–499.

Von Storch, H. and Zwiers, F.W. 2002. *Statistical analysis in climate research*. Cambridge University Press, Cambridge.

Wackernagel, M., Schulz, N.B., Deumling, D. et al. 2002. Tracking the ecological overshoot of the human economy. *PNAS Early Edition* 99: 9266–9271.

Wahba, G. 1990. *Spline models for observational data*. SIAM, Philadelphia.

Walker, B. 2002. The changing face of ecological and conservation theory. In Saunders, D. et al. (eds) *Perspectives on wildlife research: celebrating 50 years of CSIRO Wildlife and Ecology*. Surrey Beatty & Sons and CSIRO Sustainable Ecosystems, Canberra, pp. 8–13.

Wasson, R.J. 1998. Dryland farming, erosion and stream sediment: the problem of catchment scale. In Williams, J., Hook, R.A. and Gascoigne, H.L. (eds) *Farming action catchment reaction*. CSIRO Publishing, Melbourne, pp. 215–226.

WCED (World Commission on Environment and Development). 1987. *Our common future*. Oxford University Press, Oxford.

White, I. 2002. *Water management in the Mekong Delta: changes, conflicts and opportunities*. UNESCO IHP-VI, Technical Documents in Hydrology no. 61, UNESCO, Paris.

Williams, M. 1983. 'The apple of my eye': Carl Sauer and historical geography. *Journal of Historical Geography* 9(1): 1–28.

Wolman, M.G. and Miller, J.P. 1960. Magnitude and frequency of forces in geomorphic processes. *Journal of Geology* 68: 54–74.

World Bank. 1997. *Clear water, blue skies: China's environment in the new century*. World Bank, Washington DC.

Worster, D. 1994. *Nature's economy: a history of ecological ideas*. Cambridge University Press, Cambridge, 2nd edition.

Worster, Donald 1985. *Rivers of empire: water, aridity, and the growth of the American West*. Pantheon Books, New York.

Worster, Donald 1992. *Under Western skies*. Oxford University Press, New York.

Worster, Donald 1994. *Nature's economy*. Cambridge University Press, Cambridge (first edition 1977).

Yassi, A., Kjellstrom, T., de Kok, T. and Guidotti, T.L. 2001. *Basic environmental health*. Oxford University Press, New York.

Young, P. and Whitehead, P. 1975. A recursive approach to time-series analysis for multi-variate systems. In Vansteenkiste, G.C. (ed.), *Computer simulation of water resource systems*. North Holland, Amsterdam, pp. 39–52.

Index

FSC
www.fsc.org
MIX
Papier | Fördert
gute Waldnutzung
FSC® C083411

Zeitfracht Medien GmbH
Ferdinand-Jühlke-Straße 7
99095 Erfurt, Deutschland
produktsicherheit@kolibri360.de